P9-DBQ-137

Worlds Unnumbered

THE SEARCH FOR EXTRASOLAR PLANETS

This star-forming region in the constellation Serpens consists of clouds of interstellar gas and dust, within which stars and their planets are now being born, together with a host of newly made stars that have been shining for only a few million years.

Worlds Unnumbered

THE SEARCH FOR EXTRASOLAR PLANETS

Donald Goldsmith

ILLUSTRATIONS BY
Jon Lomberg

University Science Books
Sausalito, California

University Science Books
55D Gate Five Road
Sausalito, CA 94965
Fax: (415) 332-5393

Production manager: *Susanna Tadlock*
Copy editor: *Fronia Simpson*
Designer: *Robert Ishi*
Illustrator: *Jon Lomberg*
Compositor: *Wilsted & Taylor*
Printer & binder: *Maple-Vail Book Manufacturing Group*

This book is printed on acid-free paper.

Library of Congress Cataloging-in-Publication Data

Goldsmith, Donald.
Worlds unnumbered : the search for extrasolar planets / by Donald Goldsmith ; illustrations by Jon Lomberg.
p. cm.
Includes bibliographical references and index.
ISBN 0-935702-97-0 (hc)
1. Extrasolar planets. I. Title.
QB820.G25 1997 96-51826
523—dc21 CIP

Printed in the United States of America
10 9 8 7 6 5 4 3 2 1

For Rachel—
and for all
who seek new worlds

Contents

Preface

Human fascination with other worlds has a long history, resonating deeply with our desire to connect with the universe that brought us here. During the past few decades, while exploring the sun's own family of planets, astronomers have dreamt of the day when they would find what other planets circle other suns. However, the immense distances between the stars made the discovery of these planets impossible. Until now: In October 1995 astronomers found the first planet orbiting a sunlike star. The next twelve months brought half a dozen more planets to light; before the turn of the millennium, several dozen more may well be discovered.

But these planets confounded expectation. They have large masses, comparable to the mass of Jupiter, the sun's largest and most massive planet. This much astronomers expected, because the methods used to find them favor the discovery of large planets. However, half of the newfound planets orbit their stars at tiny distances, less than one-sixth of the distance from the sun to Mercury, its innermost planet. Even though the search technique also favored finding close-in planets, no one anticipated that Jupiter-like planets could exist so close to their stars as this. All theories of planet formation imply that these massive planets must have formed a hundred times farther out and somehow migrated inward to reach their present distances from their stars. This is just one of the conundrums posed by the new discoveries of extrasolar planets, which will stimulate inquiries into planet formation for years to come.

In this book, I have attempted to describe the recent planet discoveries and their manifold implications, which range over theories of planet formation, our understanding of life in the universe, the difficulties of observing extrasolar planets, and the prospects for future planet discoveries. Within a short time, extrasolar planets have made the transition from "a subject with no subject matter" to a rapidly expanding area of astronomical research, one of the most important in modern astronomy. I hope that my book captures some of the excitement, as well as the scientific facts of life (see the Glossary for terms italicized in the text) that underlie humanity's current efforts to find new worlds and thus to achieve a better understanding of our solar system and our home planet.

Donald Goldsmith
Berkeley, California
September 1996

Acknowledgments

In writing this book, I have once again had the good fortune to receive helpful assistance from many friends and colleagues, who were motivated by the desire to help me avoid the most grievous errors in describing the search for extrasolar planets. Like other authors, I hope that the suggestions and warnings from those seeking to improve the text have been properly taken into account. In any case, I hereby release those cited below from any and all liability, legal, moral, or ambiguous, in connection with the preparation of this book.

My greatest thanks go to Bruce and Kathy Armbruster, who arranged for the appearance of this book and oversaw it through the publication process; they have maintained a belief in this and other projects far beyond the requirements of business. Robert Ishi gave the book its form and design, and he and Susanna Tadlock helped me through many a difficult moment in understanding the demands of book production. Fronia Simpson copyedited the manuscript with verve. Jon Lomberg, an astronomy artist beyond easy comparison, has continued to contribute greatly to conveying scientific concepts, both in his art and by making me think more carefully about what an illustration has to say; he also made valuable comments about the manuscript. I would also like to thank Charles Beichman, Carlos Eiroa, Harley Thronson, and Jill Tarter for helping me to attend several conferences dealing with the search for extrasolar planets.

Those who provided comments, explanations, and suggestions for this work include Dana Backman, Gibor Basri, Charles

Beichman, Steven Beckwith, William Borucki, Alan Boss, Victoria Brady, Kenneth Brecher, Adam Burrows, Paul Butler, Catherine and Diego Cesarsky, Marcus Chown, Dale Cruikshank, Frank Drake, Jane Ellis, George Field, Daniel Gezari, Larry Gold, Paul Goldsmith, David Hollenbach, Michael Kaplan, David Koch, Donald Kripke, Shrinivas Kulkarni, Alain Léger, Douglas Lin, Jack Lissauer, Jonathan Lunine, Mark McCaughrean, Christopher McKay, Stephen Maran, Geoff Marcy, Lawrence Marschall, John Mather, Michel Mayor, Tsevi Mazeh, Jeffrey Moore, David Morrison, Henry Nørgaard, Donald Osterbrock, Norman Pace, Anneila Sargent, Didier Saumon, Jean Schneider, Michael Shao, Seth Shostak, Frank Shu, Michael Soule, Larry Squire, Alan Stern, Jill Tarter, Jesse Upton, Jack Welch, Michael Werner, Dan Werthimer, George Wetherill, Eric Williams, Richard Young, Ben Zuckerman, and most of all Tobias Owen, who labored long and hard to help me improve the text. To all of those mentioned here I say with thanks, without you this would have been a lesser work.

Worlds Unnumbered

THE SEARCH FOR EXTRASOLAR PLANETS

Let us (since Life can little more supply
Than just to look about us and to die)
Expatiate free o'er all this scene of Man;
A mightly maze! but not without a plan . . .

I. Say first, of God above, or Man below,
What can we reason, but from what we know?
Of Man, what see we but his station here,
From which to reason, or to which refer?
Thro' worlds unnumber'd tho' the God be known,
'Tis ours to trace him only in our own.
He, who through vast immensity can pierce,
See worlds on worlds compose one universe,
Observe how system into system runs
What other planets circle other suns,
What varied Being peoples ev'ry star,
May tell why Heav'n has made us as we are.

—Alexander Pope, *An Essay on Man* (1733)

CHAPTER

1

What New Worlds Are These?

On January 17, 1996, at the winter meeting of the American Astronomical Society in San Antonio, Texas, two astronomers from San Francisco State University, Paul Butler and Geoff Marcy, announced their discovery of two new planets, each of them orbiting a star much like our sun and (in astronomical terms) close to our own solar systems. The news fell on the astronomically inclined public like welcome rain on parched earth: At long last, after decades of searching, astronomers had found objects orbiting other stars that resemble the planets that orbit our sun. Before the month was over, Marcy and Butler found themselves named "Persons of the Week" by ABC News and featured on the cover of *Time* magazine. Their sudden prominence amazed their fellow astronomers, even though they knew that the search for planets around other stars has a special resonance with the public, since the only life we know in the universe—life on Earth—exists on a planet. Most astronomers had never dreamt that these new results would prove such a draw, elevating the search for extrasolar planets from a quiet astronomical backwater into a focus of worldwide attention.

Yet the two planets found by Butler and Marcy were *not* the first and second planets discovered around stars like the sun; instead, they were the second and third (Table 1.1). More than three months before, on October 6, 1995, in Florence, Italy, two astronomers from the University of Geneva, Michel Mayor (Figure 1) and Didier Queloz, had announced to a conference—held only a stone's throw from the home where Galileo Galilei spent years under house arrest for asserting that the Earth orbits the sun—that they had found a planet orbiting the star 51 Pegasi, forty-five light years from the solar system.

During the previous few years, Mayor and Queloz had been in friendly competition with Butler and Marcy: Each pair of astronomers had long-term observing projects that studied sunlike stars over several years' time, accumulating the data that, upon computer analysis, might reveal a planet. The astronomers had chosen to search for planets orbiting sunlike stars—stars much like our own in their masses, sizes, and surface temperatures—because the light from these stars has characteristics that make searching for planets orbiting them easier than a search around stars that are significantly hotter or cooler than the sun (see page 155).

When Butler and Marcy (Figure 2) in California learned that their Swiss competitors had found a planet orbiting a star in the constellation Pegasus, they spent a long afternoon trying to guess which star had the planet, and why they had failed to find it. Upon learning its identity, they shook their heads in dismay. The *Yale Bright Star Catalog*, a standard compilation describing several hundred thousand stars, had misclassified 51 Pegasi, listing it as significantly larger than the sun, so Butler and Marcy had dropped it from their observing list. Mayor and Queloz, who knew better, had found the first planet around another star. Electrified by the news from Florence, Butler and Marcy redoubled their efforts in observing stars and in analyzing the years of data accumulated on their computer's hard disk. Before long, they had confirmed Mayor and Queloz's discovery and had also found the two additional planets they announced in San Antonio.

Why did the initial discovery receive far less publicity than the news of two additional planets orbiting sunlike stars? A large

TABLE 1.1

Planets Found around Other Stars by Doppler-Shift Techniques, with Earth and Jupiter for Comparison

Star	*Year of Planet's Discovery*	*Star's Mass (in units of sun's mass)*	*Planet's Minimum Mass (in units of Jupiter's mass)*	*Average Planet-Star Distance (in units of the Earth-sun distance)*	*Orbital Period (days)*	*Orbital Eccentricity (see note below)*	*Distance from Solar System (light years)*
51 Pegasi	1995	1.0	0.47	0.05	4.229	0.0	45
70 Virginis	1996	0.9	6.6	0.43	116.6	0.40	72
47 Ursae Majoris	1996	1.1	2.4	2.1	1,090	0.03	44
Tau Bootis	1996	1.2	3.9	0.046	3.31	0.0	49
55 Cancri #1	1996	0.8	0.8	0.11	14.76	0	44
55 Cancri #2	1996		5	about 7	about 5,500	0	
Upsilon Andromedae	1996	1.2	0.6	0.04	4.6	0	54
16 Cygni B	1996	1.0	1.6	2.2	810	0.57	84
Sun (& Earth)		1.0	0.003	1.0	365.24	0.02	
Sun (& Jupiter)		1.0	1.0	5.2	4332.6	0.05	

Note: For a planet's elliptical orbit around a star, the orbital eccentricity equals the difference between the planet's greatest and least distances from the star, divided by the length of the ellipse's long axis. For a circular orbit, this ratio is zero; for a tremendously elongated ellipse, the eccentricity approaches one.

FIGURE 1 Michel Mayor of the University of Geneva is the codiscoverer of the first planet found around a sunlike star.

part of the explanation for the difference lies in Europe's distance from the United States, but some of it resides in a crucial fact about science. The most basic rules for scientific investigation of the universe imply that the discoveries of the second and third examples have greater significance than the first. *One* planet, discovered around *one* star beyond the sun, might prove a fluke, perhaps not a planet at all, perhaps explicable by stellar pulsations or by some other phenomenon only dimly understood. But when several stars display similar patterns of behavior; when that behavior can be most straightforwardly attributed to the effects of a planet moving in orbit; and when the planets deduced to exist from these effects (see page 9) are roughly similar yet differ in their masses and distances from their parent stars—then scientists can reasonably conclude that they have met a widespread phenomenon, not some fascinating but immensely unusual freak of nature. Before 1996 had ended, a number of newfound planets had risen to the point that astronomers could almost begin sta-

FIGURE 2 Geoff Marcy (left) and Paul Butler (right) found six of the first seven planets around stars similar to our sun.

tistical analysis of the new planets' properties—an impossibility with only a single example.

In 1733 the poet Alexander Pope recognized this principle in his famous poem *An Essay on Man*, from which an excerpt appears as this book's epigraph, by asking, "Of Man, what see we but his station here, / From which to reason, or to which refer?" Pope had hit upon the utterly valid point—often forgotten by philosophers, politicians, and poets with axes to grind—that we can draw conclusions only from what we know, but when the day arrives that we know more, why then our conclusions may change. When, a few lines later, Pope speculated about planets circling other stars, he wrote of "worlds unnumber'd." This phrase, which

I have chosen for the title of this book, underscores (at least for me) the difference between what we humans know today and what we may yet know and enumerate in the future.

Today, having numbered most of the stars within our view, we are beginning to number the worlds of which Pope dreamt. This occurs almost literally, since the stars with newly discovered planets typically have catalogue numbers such as 70 Virginis and 51 Pegasi, which refer to a list of the stars within a particular constellation. Because astronomers lovingly cling to a nomenclature system first developed during Pope's lifetime, the constellation name appears in Latin and in the genitive (possessive) case, so that 70 Virginis is star number 70 in the constellation Virgo the Virgin. The brightest individual stars in each constellation are designated by letters of the Greek alphabet: Aldebaran is Alpha Tauri, Antares is Alpha Scorpionis, and the seven stars of the Big Dipper are Alpha through Eta Ursae Majoris.

What Do the New Planets Mean to Us?

Until 1995, amid the entire host of stars that spangle the night skies, we knew just one example of planets orbiting a star: our own solar system. Despite astronomers' steadily increasing knowledge of the sun's nine planets, of the planets' sixty-odd satellites, and of a host of sun-circling asteroids, comets, and meteoroids, we had no knowledge of planets around other stars, save for the fact that detecting any such planets presents an extremely difficult challenge. In this situation, astronomers did what comes naturally. They extrapolated from what we know about the present and past of the solar system to draw conclusions about planetary systems that might accompany stars other than the sun. Inevitably, despite their disclaimers about the unreliability of building an empire of speculation on a single example, astronomers tended to use our solar system as the representative model of what to expect around another star: smaller, rocky planets close in, giant planets farther out, with small objects moving among the planets in relatively odd orbits, occasionally colliding, sometimes with catastrophic results.

The first three new planets changed astronomers' conclusions about where to expect giant planets. As has often occurred in the annals of scientific discovery, the new objects turned out to be odder than speculation had predicted—just odd enough to call into question much of what astronomers thought they understood about how planets form. The first planet found, in orbit around the star 51 Pegasi, began this upset: Although its mass, at least half that of Jupiter's, marks it as a giant planet, this planet orbits its star once every 4.2 days, at a distance less than one-seventh of Mercury's distance from the sun (Color Plate 1)! Mercury, the innermost planet in our solar system, orbits the sun at 38 percent of the average Earth-sun distance. This distance, called an *astronomical unit* and denoted as 1 A.U., is slightly less than 93 million miles or 150 million kilometers. The planet around 51 Pegasi orbits at only 0.05 A.U. from its star—just 1/20 of the Earth's distance from the sun, and so close that heat from the star must raise the planet's temperature well above 1,000 degrees Celsius. No astronomer expected to find Jupiter-like planets so close to a star. Theoretical calculations had shown that as the solar system formed, ice—the key to making giant planets—could form in abundance only at distances at least several times greater than the Earth-sun distance. These theories, together with the evidence from the solar system, had led astronomers to expect giant planets like Jupiter to be discovered (should they exist) at distances of at least 4 A.U. from the star.

Once found, the planet cried out for explanation—an explanation that astronomers have provided, with serious implications for the possibilities of finding Earth-like planets (see page 158). Meanwhile, the planets around two other stars gave astronomers some relief, for they appeared in more familiar orientations—but these planets raised problems of their own for those who thought they had planetary formation down cold. Of these two planets, the stranger one orbits the star 70 Virginis, about 72 light years from the solar system. This planet has a mass at least 6.6 times

Jupiter's, which might give it enough mass to be a brown dwarf (see page 88). It orbits its star once every 116.7 days, about half the Earth-sun distance, though it moves in a noticeably elongated orbit, varying its distance from the star by more than a factor of two, from 0.27 to 0.59 A.U. (Color Plate 2). Just as astronomers must strain to explain a high-mass planet orbiting close to its star, so too must they exert themselves mightily to understand how a massive planet can have such an elongated orbit. In the solar system, Jupiter and the other giant planets move in nearly circular orbits, and only relatively tiny objects such as comets and asteroids have orbits as elongated as the one found around 70 Virginis.

The third new planet, found 44 light years away, orbiting the star 47 Ursae Majoris, most nearly corresponds to what astronomers expected. This planet, with a minimum mass of 2.4 Jupiter masses, has an orbit somewhat larger than Mars's orbit around the sun, and takes 1,090 days—just under 3 years—for each revolution (Color Plate 3). Hence the planet around 47 Ursae Majoris most nearly fulfills astronomers' expectations as to the size and shape of its orbit, though it too lies closer to its star, for a planet with a mass similar to Jupiter's, than astronomers can easily explain.

How Did Astronomers Find These Planets?

The first planets to be found around other stars have never been seen; most of them are not likely to be seen for many decades. Planets nestle so close to their parent stars, and shine so weakly in comparison, that astronomers have little chance of *seeing* Jupiter-sized planets around even the closest stars. The search becomes much more difficult as we seek Jupiter-like planets that orbit closer to their stars than Jupiter does to the sun, which is in fact the situation for all the planets discovered in 1995 and 1996. To find these planets, astronomers relied on one of the most useful and best-tested laws of physics: the Doppler effect.

Long valued as a highly precise and informative instrument

in the astronomer's toolbox, the Doppler effect describes the fact that anyone who observes wave motion will detect a change in the waves' frequency (the number of wave crests arriving each second) and their wavelength (the distance between successive crests) whenever the source of waves moves toward or away from the observer, or whenever the observer moves toward or away from the source (Color Plate 4). Motion toward the observer crowds the waves together, so the waves' frequency rises and their wavelength decreases. In contrast, the recessional motion of the source stretches the intervals between crests, so the frequency falls and the wavelength increases. More rapid motion produces greater changes, and the effect does not depend on whether the source of the waves moves, the observer moves, or both move. In any case, an observer will find wave crests from an approaching source arriving more often, and those from a receding source less often, than from a source at rest with respect to an observer. The greater the relative velocity of approach or recession, the greater will be the changes produced by the Doppler effect.

To our eyes, the frequency and wavelength of light waves appear as *colors*. For all types of light, the product of frequency and wavelength remains the same: the speed of light, denoted by **c** and equal to about 186,000 miles per second. Thus waves with larger frequencies have smaller wavelengths, and vice versa. Of all the light our eyes can detect, red light has the longest wavelengths and smallest frequencies, whereas violet has the shortest wavelengths and highest frequencies. The full span of visible-light waves, from smallest to largest, amounts to a factor of about two in wavelength and frequency. An observer approaching a red light traveling at about 80,000 miles per second (nearly half the speed of light) would see the light's color as green—not a good story to tell the judge. On either side of the visible-light span lie other portions of the spectrum of electromagnetic waves, or electromagnetic radiation. At frequencies below those of visible light, we find radio, microwaves, and infrared; at higher ones, we encounter ultraviolet, x rays, and gamma rays. Although astronomers have developed techniques to study nearly the entire spectrum of this radiation, the new planets revealed themselves through the Doppler effect acting on the most familiar part of this spectrum, visible light.

Astronomers had long used the Doppler effect to find fainter stars moving in orbit around brighter ones. In many of these double-star systems, the brighter star's light so overwhelms the fainter one's that astronomers can observe only a single spectrum, that of the light of the brighter star. Nevertheless, by measuring the frequencies and wavelengths of the features in that star's spectrum, they can often detect periodic, repetitive variations, first toward the red and then toward the violet ends of the visible-light spectrum. From these cyclical changes, astronomers can deduce the existence of a second object, whose gravitational force pulls the star under observation first in one direction and then in the opposite direction as the two stars orbit their common center of mass.

In a similar way, the Doppler effect can reveal planets orbiting a star—provided that the planets' gravitational forces produce detectable changes in the features observed in the star's spectrum. Planets orbit their stars because those stars exert gravitational forces on them; without these forces, the planets would long since have escaped into the depths of space. As Isaac Newton explained, any object in motion, as all planets are, will move in a straight line if no net force acts upon it. Since all planets are born in motion, they must all be "held" by their stars' gravitational forces, to which they respond by moving in closed orbits, along every point of which gravity counteracts the planet's tendency to sail off in straight-line motion, and the motion counteracts gravity's tendency to pull the planet directly inward. If we could observe planets around other stars by the light they reflect from their stars, we would also expect to see changes in the frequency and wavelength of that light: changes that result from the planet's motion and described by the Doppler effect.

But *we can't see planets*. How, then, can astronomers find them by means of the Doppler effect? The answer lies in the universality of gravity. Every planet attracts its star—with the same amount of gravitational force that the star exerts on the planet. These two forces provide a fine example of another Newtonian discovery, which many of us learned in high school, that every action has an equal and opposite reaction. In modern terminology, the *forces* that each object exerts on the other are equal in strength, but the two objects *respond* to those forces quite differently. Newton tells us that any object will accelerate in response to a force, but *the amount of acceleration varies in proportion to one over the object's mass.* When a golfer "tees it high and lets it fly," the ball accelerates in an instant from zero to a velocity of several hundred feet per second. However, if an inebriated golfer should reach into his capacious bag and tee up a baseball, the acceleration from the same amount of force will be far less—only 1/100 as much if the baseball has 100 times the golf ball's mass.

Comparing the drunken golfer to the planet and star illustrates that a star does accelerate in response to its planet's gravitational force, but far less than the planet accelerates in response to the equal amount of gravitational force from the star. Both the star and planet perform orbits around the *center of mass* of the planet-star system, a point that always lies on the line joining the centers of the two objects. The distances from the center of mass to the two objects' centers are inversely proportional to the objects' masses. If a star has a mass 10,000 times greater than its planet, the center of mass lies 10,000 times farther from the planet's center than from the star's center. This puts the center of mass inside the star, but not quite at its center. As the planet moves in orbit around the center of mass, the star too performs its own, much smaller orbit around the same point, like a decayed nobleman making small, slow-moving imitations of much younger athletes' pirouettes and jetés.

This analogy has its accurate side: Planets in orbit around other stars move at speeds of many kilometers per second, but their stars move at only a few meters per second—the speed of an Olympic athlete. In these few meters per second we find the current best means for discovering other planets (see chapter 6 for a discussion of other promising techniques). Suppose, for example, that a sunlike star has a planet moving at 10 kilometers per second in orbit around it. If this planet has 1/1,000 of the star's mass (Jupiter, for example, has 1/1,000 of the sun's mass), then the star will move at 10 meters per second. The task facing an astronomer hoping to find planets then amounts to detecting the changes described by the Doppler effect, called *Doppler shifts*, which arise from motions at speeds of a few meters per second. These Doppler shifts in the spectrum alternately shift the features in the spectrum of starlight toward higher and lower frequencies. If the changes repeat in a definite cycle, the astronomer may reasonably conclude that something tugs the star, first in one direction and then in the opposite direction (see Figure 1.6). The time interval for the cycle to repeat itself reveals the orbital period of the object tugging the star, and some sophisticated analysis will reveal the object's mass, and even the deviation from circularity of its orbit.

Why, then, did it take so long for the Doppler effect to reveal extrasolar planets? The challenge of achieving an accuracy of a few meters per second in Doppler-shift measurements required several decades of hard work. For light waves, the fractional change in frequency and wavelength produced by the Doppler effect varies in proportion to v/c, where v is the speed of the object and c is the speed of light, close to 300,000 kilometers per second. Velocities of 3 meters per second (0.003 kilometer per second) will therefore produce fractional changes equal to (0.003/300,000), or one part in a hundred million. This level of precision lay beyond astronomers' abilities until a few years ago.

In fact, the work by Marcy and Butler in California and by Mayor and Queloz in France (for the Swiss astronomers observed stars from the Haute Provence Observatory in the Pyrenees) found the first planets around other stars just after the previous best search had been abandoned as fruitless. For a decade, starting in the early 1980s, the Canadian astronomer Bruce Campbell had sunk all his astronomical efforts into a Doppler-shift survey of twenty-one nearby sunlike stars. "Bruce was the grandfather of this effort," Marcy says. "He was the stick that poked our sides." But Campbell's spectrometer—the instrument used to measure frequencies and wavelengths—could not quite reach the accuracy that proved necessary for planetary discovery. All accurate spectrometric measurements employ an instrument (formerly a prism, but now usually a reflection grating ruled with finely spaced parallel lines) that spreads light into its component wavelengths and frequencies, ranging from the red to the violet end of the visible-light spectrum. The instrument can also produce a comparison spectrum when white light (that is, light with all frequencies and wavelengths present in equal amounts) shines through a known substance into the spectrometer. The known substance removes light with particular colors (that is, with particular wavelengths and frequencies), thus producing dark *absorption lines* in the spectrum. The spectrum of light from an astronomical object likewise typically contains numerous absorption lines, but it may also show bright *emission lines*. Emission lines have those frequencies and wavelengths for which atomic and molecular processes produce large amounts of light, whereas absorption lines appear at the frequencies and wavelengths at which atoms and molecules block light. Once the frequencies and wavelengths in the comparison spectrum have been carefully measured, astronomers can use them for their designated purpose: comparison with the frequencies and wavelengths of the emission and absorption lines in the spectrum of the astronomical object under study.

The choice of a source for the comparison spectrum proved fateful. Campbell and many other observers had favored the spectrum produced when white light passes through hydrogen fluoride (HF) gas. This produces a long-studied, well-understood spectrum, rich in absorption lines. Though the HF comparison spectrum covered only a small portion of the total spectrum of visible light, its absorption lines are conveniently spaced for comparison with lines in the stellar spectra. However, in 1992, by the time that Campbell and his collaborator, Gordon Walker, had suspended their efforts, eventually producing a scientific paper entitled "On the Absence of Jupiter-like Planets around Solar-

Type Stars," Marcy and Butler had put their money on a different comparison spectrum: the absorption-line spectrum produced when white light shines through iodine gas.

Marcy, now in his early forties, trained as an astronomer with a particular interest in low-mass stars. After finishing graduate school at the University of California at Santa Cruz, he obtained a Carnegie Fellowship, which allowed him to engage in unsupervised postdoctoral work at the Carnegie Observatories, endowed by Andrew Carnegie soon after the turn of the century and located on Santa Barbara Street in Pasadena, California; astronomers there often refer to the observatories as "Santa Barbara Street" to distinguish that location from "the mountain," denoting the giant telescopes on either Mount Wilson or Palomar Mountain. ("There is *no* Mount Palomar," older astronomers will insist if you lapse into momentary error.) At Santa Barbara Street, Marcy soon found himself deeply depressed. "I'd already felt less than mediocre as a graduate student," he notes, "and when I got my 'prestigious Carnegie Fellowship,' that feeling was heightened. It seemed as though other fellows knew a hundred times more astronomy, and I didn't have the 'right stuff.' "

Before long, Marcy was in psychotherapy. One day he found himself unable to go to work; instead, he stood in the shower, determined not to emerge until he had found a project that he could call his own and that would allow him to hold his head up at Santa Barbara Street. Eventually, he hit on the notion of searching for *brown dwarfs*, objects a bit too small to become stars, which we shall meet again in chapter 4. This was a project he could wrap his mind around and into which he enlisted the collaboration of the most famous of the Santa Barbara Street astronomers, Allan Sandage, an expert on both stars and galaxies, each of which, like our Milky Way, contains millions or billions of stars.

Profiting from the astronomical endowment of Santa Barbara Street, Sandage and Marcy shared 8 nights per month observing with the 100-inch telescope on Mount Wilson, whose location close to Los Angeles had left it no longer at the peak of astronomers' desires. With this great instrument, which Edwin Hubble had used during the 1920s to discover the expansion of the universe, they studied 70 nearby stars, using Doppler-shift techniques to search for possible brown-dwarf companions. Because the 100-inch telescope had a relatively antiquated spectrometer, they could hope to detect only objects that made their stars move at speeds of at least 230 meters per second; these were likely to be objects with masses 10 to 100 times greater than Jupiter's mass, hence brown dwarfs. "We made the first absolutely definitive *non*detection of brown dwarfs," Marcy recalls. Other astronomers, aware of his results, called him "Doctor Death." "I took that as a badge of honor," says Marcy, "since in some ways a *non*detection has more integrity than a marginal [not truly definitive, or, your opponents would say, not at all definitive] detection."

In 1984 Marcy obtained a faculty position and moved from Santa Barbara Street to San Francisco State University, a journey that in astronomical terms resembles a baseball player's move from the Atlanta Braves to the Birmingham Barons. He still traveled to Southern California for occasional nights of observing with the 100-inch telescope, and he continued the relationship with the astronomers in charge of the Lick Observatory near San Jose, whose 120-inch reflector took pride of place within the University of California system, and which Marcy had used for his thesis research. The Lick astronomers gave Marcy a few nights per year to continue his search for brown dwarfs, using the existing spectrometer, with a velocity resolution of about 200 kilometers per second. When Marcy read Campbell's first published paper, he saw that he could achieve higher accuracy if he switched to a hydrogen-fluoride system for the comparison spectrum. "But HF is invisible, it's toxic, it seemed like trouble."

At about that time, near the end of 1984, Paul Butler, a graduate student with a chemistry background, had arrived at San Francisco State; he introduced himself to Marcy, who thought that Butler might be the person to help him find a source for a more accurate comparison spectrum. "I went into Paul's office," Marcy says, "and proposed that we make our own absorption cell [to produce a comparison spectrum], one not as toxic as HF, and one that would cover more wavelength." The HF spectrum covered only about 1 part in 400 of the visible-light spectrum. This made comparison with the part of the stellar spectrum in that range simple, but Marcy recognized that a wider spread for the comparison spectrum would produce greater accuracy, even though the comparison process would become more difficult. Butler knew something about the different types of absorption cells that produce a comparison spectrum dominated by absorption lines, for they find widespread use in many areas of applied chemistry—not a subject on which most astronomers are expert.

"We went to great pains," Marcy recalls. "We took six months looking for the right substance; we looked up laboratory spectra; and I wrote to Herzberg. *The* Herzberg." Gerhard Herzberg, a German-born Nobel Prize winner, was for decades a professor and researcher in Canada and the author of fundamental texts on spectroscopy. "And Herzberg said: 'Iodine.'" The man clearly was not wedded to any one system, since it was he who had recommended hydrogen fluoride to Bruce Campbell. Iodine gas as the source of a comparison spectrum offered advantages and disadvantages: It produced an immense number of absorption lines, but for that reason alone, comparing the iodine spectrum with that from an astronomical object was a daunting task, even with the latest computers. "The solar astronomers at Mount Wilson had dabbled with iodine," Marcy remembers, "but they didn't want to do the sophisticated analysis to model the forest of iodine lines." In fact, iodine produced so many lines that they tended to "contaminate"—overlap with—the spectral lines produced in the object under study. "But we said, So what? This is the computer age! We ought to be able to make models that compute the combined effect of both the comparison and the star. That was a leap—and it paid off."

In 1987 Marcy and Butler deployed a system at Lick Observatory that used a comparison spectrum from iodine to search for Doppler shifts in the light from nearby stars. Relying on small

grants from the National Science Foundation, they persevered; just eight years later, they had found their first planets. "We made nine years of mistakes," Marcy calculates, figuring in the time it took to develop the iodine system. And despite his So what? attitude, he cheerily admits that "no one had thought through how to carry out the analysis. I remember thinking, this could be a big embarrassment."

To make the most precise possible measurements of their comparison spectrum, Marcy and Butler took their iodine system to the giant solar telescope at the Kitt Peak National Observatory in Arizona. This produced what Paul Butler calls, with only a touch of irony, the "sacred iodine spectrum." They eventually realized that their comparison of the multitude of absorption lines in the comparison spectrum with the lines in the star's spectrum had to include computer models of the ways that the spectrometer spread out the light from a star, by different amounts at different wavelengths. A hardworking graduate student, Eric Williams, now at Wesleyan University, worked with Marcy and Butler to obtain most of the observations and to perform the analysis that yielded the second and third planets.

By 1993, two years after Bruce Campbell had announced that he was abandoning his research (he is now a businessman in Victoria, British Columbia), Marcy and Butler were moving toward a system that could achieve a velocity resolution of three meters per second, which they have now achieved, profiting from key assistance from Lick Observatory's expert on spectrometry, Steven Vogt. During that same year, Mayor and Queloz had developed a spectrometer that uses fiber optics to bring the comparison and stellar spectra to a location away from the telescope for comparison, rather than superimposing the comparison spectrum directly on the stellar one, so that a single exposure records both simultaneously. In this system, the comparison spectrum consists of the emission lines produced by heating the element thorium. Although their system has a greater likelihood of developing mechanical difficulties, Mayor and Queloz overcame all such problems to produce a spectrometer that can measure velocity changes as small as thirteen kilometers per second, quite sufficient to discover the planet around 51 Pegasi.

Why were Marcy and Butler "so slow?" One reason is that until Mayor and Queloz made their announcement, astronomers expected to find Jupiter-like planets, should any such exist, moving in Jupiter-like orbits. The Doppler-shift method reveals the most massive planets most easily, because those planets produce the largest effects on their stars. Furthermore, the closer such a planet orbits to its star, the larger will be the effect that it produces, since its gravitational force on the star will be greater. But the single example of our own solar system had persuaded astronomers (or so it seemed—many of them later explained that they had never been so fully persuaded) that Jupiter-like planets must move in large orbits, taking at least a few years per orbit. To detect such a planet would require at least a couple of orbits, in order that astronomers might convince themselves that they had found a repetitive phenomenon, not a onetime event. Thus the astronomers' plans ran toward accumulating a decade or so of data, then analyzing it carefully, searching thoroughly for planets moving in ten- or twenty-year orbits.

By the late summer of 1995, although Marcy and Butler had analyzed the data for only about 20 of the 120 stars on their observation list, they felt no great pressure to pick up the pace. They lacked access to truly fast computers, and their attempts to secure research grants had met with only modest success. Scientists compete for governmental support against one another, with every grant application receiving anonymous reviews from several referees. NASA had made a significant grant (by Marcy's standards) in 1992, but in August 1995, Marcy received a report from the panel to which NASA headquarters had referred his most recent grant application. The panel stated that "we are . . . sure that [Marcy] realizes that the radial velocity study of a star can not [quoting from Marcy's application] 'provide the first definitive evidence regarding the existence (or absence) of Jupiter-like planets' in orbit around any individual star . . . the likelihood of detection should be presented in statistical terms." As Marcy wryly notes, "It's no fun to be accused of deliberate deception." The panel also pronounced itself "unconvinced that another factor of 2 or 3 in precision could be obtained" and noted that "[t]he long-term performance of the instrument is still unpublished. . . . To the extent that the program can detect planets, relevance is high, but in view of the panel's reservations about the feasibility of extending accuracy . . . relevance to [NASA's] Origins [project] may be limited to searching for infrared ('brown') dwarfs." An accompanying letter from NASA informed Marcy that his proposal had received the most minimal level of funding.

Two months later, Mayor and Queloz's announcement of the planet around 51 Pegasi quickly increased the perceived relevance of Marcy and Butler's search. It also significantly altered astronomers' notions of the orbits of planets that might be discovered. The planet around 51 Pegasi orbits so close to its star that it takes not years, not months, but *days*—4.2 of them—to complete each orbit. This fact produces two effects that made the planet easier to find. First, the smaller distance means that the planet exerts more force on the star, thus inducing more rapid motion of 51 Pegasi. Still more important, it also means that even a year contains more than enough data to reveal the planet's existence unequivocally, which could never be the case for a planet in a multiyear orbit. On October 6, 1995, when Marcy and Butler heard about the new planet, they had already obtained observing time on the 120-inch telescope at Lick Observatory, scheduled for the nights of October 11–14. "Four nights!" Marcy recalls, "and the [orbital] period was four days! And they were all clear! We worked out a three-ring circus: Eric Williams was at San Francisco State to run the computer program removing the Earth's motion from the analysis, while Paul Butler and I stayed at Lick reducing the data. By the second night, we saw that the velocity had changed; by the third night, we saw the change in the velocity change—the star had an orbit."

From the *observation* of this stellar orbit (see Color Plate 1) came the *deduction* of the planet's existence. The star 51 Pegasi showed cyclical changes in velocity, repeating over a 4.2-day cycle, exactly what would be expected if a planet circled the star with a 4.2-day orbital period. The ingrained skepticism of scientists caused Marcy and Butler—along with Mayor and Queloz and every other astronomer who dealt in the realm of extrasolar planets—to ask what other effects could produce the observed changes in the star's velocity. In their paper announcing the discovery, Mayor and Queloz ran through the possibilities. Could the star be pulsating, moving its surface outward in a cyclical manner? This explanation seemed far less likely than motion caused by an object orbiting the star, because no sunlike star had been observed to pulsate with such stunning regularity, with the same changes in velocity in each cycle. Could 51 Pegasi have dark spots and could its rapid rotation mimic the changes in the spectrum produced by the Doppler effect? Almost no stars rotate so rapidly—none of them sunlike—and to mimic Doppler shifts seems nearly impossible. Furthermore, Mayor and Queloz actually determined the rotation of 51 Pegasi, which broadens all the lines in the star's spectrum by a measurable amount, again thanks to the Doppler effect. The star's rotation period turned out to be 30 days, almost the same as the sun's and many times greater than the 4.2-day cycle in the spectral lines. Other alternative possibilities appeared still more remote, and the news went forth: 51 Pegasi has a planet.

And what a planet! Astronomers knew that the Doppler-shift method of search would find the most massive planets first and would reveal closer-in sooner than farther-out planets. But they never expected to find a planet with a mass similar to Jupiter's so close to its star: just 1/100 of the Jupiter-sun distance. How could this planet have grown so large? According to the accepted picture of planet formation, the star's heat would prevent the formation of a massive, ice-rich core that would allow a giant planet to form.

Chapter 4 deals with this weighty issue by considering how planets can form, and how they can migrate from orbit to orbit during their early history. For now, though, we must consider a key subtlety connected with the astronomers' conclusions about the *masses* of the planets found around other stars. Whenever astronomers deduce the existence of an object in orbit around a star by observing velocity changes in the light from the star, they immediately obtain an accurate measurement of the object's orbital period (the time for the changes to pass through one cycle) and of the deviation of the orbit from being perfectly circular (noncircular orbits, in which the object moves more rapidly at some points along the orbit than at others, produce deviations from symmetry in the velocity changes). But the mass deduced for the object represents only the minimum possible mass that the object can have. The Doppler effect reveals only the changes in the star's velocity along our line of sight and does not provide information about the star's motion in the perpendicular directions. This means that unless the plane of the planet's orbit happens to coincide exactly with our line of sight, we observe only *part* of the star's motion (Color Plate 5). Suppose instead that the plane of the orbit is exactly perpendicular to our line of sight. Then we observe no Doppler shifts, no matter how rapidly the star may be moving in response to its planet's gravitational force. Intermediate cases produce Doppler shifts of intermediate sizes. If, for example, the plane tilts by 60 degrees away from our line of sight, we observe a velocity component that is only half of the true velocity: If the star actually moves at 10 meters per second in response to its planet's gravitational force, we observe only 5 meters per second of that motion.

Since the deduction of the planet's orbital period and orbital eccentricity (deviation from circularity) are not affected by the tilt of the orbital plane with respect to our line of sight, we can rely on those two numbers. But in nearly every case we shall underestimate the planet's mass if we assume that the Doppler effect reveals all of the star's motion. If we derive a mass for the planet based on a particular orbital period and an observed stellar velocity of 5 meters per second, the derived mass will be just half of the actual mass in the case that the total stellar velocity equals 10 meters per second, that is, if the orbital plane tilts at 60 degrees with respect to our line of sight. Hence the masses listed in Table 1.1 are rock-bottom minima.

How large could the *actual* planetary masses be? Here we meet the importance of discovering many (or at least several) planets, not just one. In October 1995, when astronomers heard the news of a "planet" around 51 Pegasi with "at least half" of Jupiter's mass, they all knew that Mayor and Queloz might have found a low-mass companion *star* for 51 Pegasi, moving in an orbit so nearly perpendicular to our line of sight that the Doppler effect reveals only, say, half a percent of the star's actual velocity. Then the object with at least half of Jupiter's mass might in fact have 100 times Jupiter's mass, or 1/10 of the sun's mass, making it a small star or a brown dwarf (see page 88), but not a planet.

Nature may be subtle, but is not malicious, as Einstein once put it. When astronomers search for Doppler shifts large enough to detect, they may well find one example of a double-star system aligned by chance to mimic a planet. But when they find several such systems and (here comes the clincher) *find no systems with higher minimum masses for the orbiting objects*, they may rightly conclude that they have discovered a new class of objects, not stars in oddly aligned orbits, but planets. The searches by Mayor and Queloz, by Marcy and Butler, by Campbell and Walker, and by other astronomers engaged in similar projects, could reveal objects with masses between 20 and 100 Jupiter masses, provided that the objects orbit relatively close to their stars, and that their orbital planes are not nearly perpendicular to the line of sight. But no such objects had been found. (We shall discuss the exceptions to this statement, which do not change the conclusions, in chapter 4.) Because we expect that the distribution of orbital planes with respect to our lines of sight to be random, and because astronomers found almost none of the higher-mass orbiting objects, their discovery of a new *class* of objects with minimum masses between 0.47 and 6.6 Jupiter masses stands on a secure basis, not explicable as the result of a chance alignment of the orbital plane with respect to our line of sight.

For purposes of feeling secure in the preceding argument, we may ask, How many objects constitute a class? When I was a boy, interested in mathematics, my father fascinated me with tales of tribes in Brazil whose system of counting was "one, two, three, many." No concept of separate numbers beyond three existed, presumably because the tribes had no need of them. I believe such people actually existed; at any rate, their example proves useful for scientific purposes. With eight examples of planets in the low mass category (0.5–6.5 Jupiter masses), we have strong, though not yet utterly compelling, evidence that the low-mass category forms a separate class, even if one or two of the highest-mass objects among the eight turn out to be brown dwarfs.

Astronomers quickly adjusted to the news that some nearby sunlike stars do have planets. Several significant questions immediately arose: What fraction of all sunlike stars have planets? How many planets orbit each star? What are those planets like? The answers to these questions must await further discoveries, which will permit proper statistical analysis. Before reaching this happy ground—even before indulging in speculation based on the first planetary discoveries—astronomers had to examine the greatest single issue raised by the first planets discovered around other stars: How did most of them come to orbit so close to their stars?

Five of the first eight planets raise this issue in spades. As Table 1.1 shows, these five orbit their stars at distances less than 1 A.U.—just under half an A.U. for the planet around 70 Virginis, just over one-tenth of an A.U. for the inner planet around 55 Cancri, and mind-bending distances of 0.05 A.U., 0.046 A.U., and 0.04 A.U. for the planets orbiting 51 Pegasi, Tau Bootis, and Upsilon Andromedae, respectively. One-twentieth of an astronomical unit! No astronomer had suggested that a planet with a mass like Jupiter's could be found so close to its star. And yet, before 1996 had advanced beyond the bite of winter, many astronomers had concluded that finding giant planets in such proximity to their stars made perfectly good sense. Because science saves its greatest praise for the person who uses cogent analysis to predict strange phenomena *before* they are found, we may feel sure that some theorists must regret their failure to perceive, during the 1980s and early 1990s, just how reasonable their explanations for giant planets orbiting at 0.05 A.U. from their stars would prove to be.

Energized by the actual discovery of planets, theorists have risen to the challenges posed by the small planet-star distances. Before examining the explanations they have produced, we must deal with another fascinating aspect of the new discoveries: Could these planets harbor life?

CHAPTER

2

Could the New Planets Harbor Life?

When the news of planets discovered around other stars burst onto an unsuspecting world, the question was immediately asked, How suitable are these planets for life? Astronomers, quite aware that the discovery of life on another world would mark a key moment in human history, had anticipated this inquiry. The key moment may have occurred on August 7, 1996, when researchers announced intriguing evidence for life in a meteorite from Mars. If this announcement should turn out to have been a false alarm, the new planets may yet prove to provide the first known sites for extraterrestrial life. Even if the first discovery of extraterrestrial life occurs within our own solar system, the possibility of life on these other worlds circling other stars remains fascinating both to scientists and to the wider public.

To inquiries about life on the new planets, astronomers responded from years of experience in research and argumentative explanation. Most of the answer that came through to the public focused on temperatures and stated that the "close-in" planets, such as those around 51 Pegasi and 70 Virginis, are almost certainly too hot for life to exist. On the other hand, the planet around 47 Ursae Majoris, orbiting more than twice as far from its star than Earth does from the sun, *might* have temperatures that allow life to exist. However, life seems to require significantly more than the proper temperatures. To understand and to assess these pronouncements, we must explore what was once called a "science without a subject matter": the nature of life beyond the Earth.

The Origin and Development of Life on Other Planets

All that we know from our investigation of life on Earth, and our attempts to apply those results to the possibilities of life elsewhere, suggests that living creatures consist of cosmically abundant elements and that life arose through natural processes, evolving by natural selection to produce the fantastic efflorescence of life forms that we see today, nearly four billion years after life began. These facts imply that wherever Earth-like raw materials and environments may exist, so too should life, and that those forms of life should likewise experience natural selection—the struggle for reproductive success that leads to new types of organisms.

But the devil resides in the details. Just how Earth-like must conditions be to produce living creatures? How might they resemble, and differ from, those on our planet? For now, the fact that we know only a single form of life—life on Earth—cruelly handicaps our ability to draw firm conclusions. A sample set of one cannot yield good statistical analysis. The discovery of even one more type of life—fossil life on Mars, life beneath Europa's icy crust, life in pools of liquid hydrocarbons on Saturn's moon Titan, or life on a planet around 70 Virginis—would tremendously increase our analytical powers through the comparisons we could make between the biochemistry and metabolic processes of Earth-born and other-born life.

These comparisons of similarities and differences between forms of life would allow us, for instance, to draw strong conclusions about the panspermia hypothesis, the possibility that life has somehow traveled through interstellar space, passing from site to site throughout the Milky Way. In addition, we could address a host of questions that demand definitive answers. Do planets provide the most favorable places for life to originate and to evolve? We think so. Do all forms of life rely on carbon atoms for their basic chemistry? We think they do. Do they all require a liquid solvent, in which the molecules that participate in life's processes can float and interact? We expect so. Does this mean that life should exist primarily, or exclusively, in regions where the solvent remains liquid? The conclusion follows naturally from the need for a solvent, but we may underestimate life's ability to persist, perhaps in a form of hibernation, under conditions that appear to preclude its continuation. Such a short paragraph, so much uncertainty! Let's take a few moments to run through the generalizations above and see why biologists believe that they qualify as valid.

Life on Planets and the Superiority of Carbon

First, consider the advantages that planets offer for life. If we define *life* as molecular systems that possess the capacities to reproduce and to evolve, we can admire planets for the fact that they offer an excellent mix of ingredients to create molecular mixtures, plus the best conditions in the cosmos for these molecules to interact. This mix of potential life-making ingredients builds from the basic units of matter, the atoms that can combine to form molecules.

Of all the varied types of atoms that exist in the universe, carbon atoms have the greatest ability to form complex molecules when they combine with other atoms. Why? Because carbon atoms can form bonds with one, two, three, or four different atoms, a flexibility lacking among most other atomic species. Each hydrogen atom, for instance, can bond with only one other atom at a time. Oxygen atoms can combine with one or two atoms, but not with three or four, and nitrogen atoms can bond with one, two, or three atoms, but not with four. The reasons for these differences lie in the nature of atoms, described by the quantum-mechanical rules that govern how the electrons in atoms orbit the central nucleus; they make carbon nature's favorite atom in its ability to form chemical bonds with up to four other atoms at a time.

Carbon's ability to bond with as many as four other atoms finds a match among silicon atoms, the favorite in science-fiction novels as an alternative basis for the chemistry of extraterrestrial life. To be sure, silicon can match carbon in numerical bonding ability, but the bond between two silicon atoms has only about half the strength of the carbon-carbon bond. Life on Earth relies on an enormous variety of atoms with a carbon backbone, a long chain of carbon atoms bound to each other, with other types of atoms attached to the sides of the chain. Analogous complex molecules with silicon backbones would be considerably more fragile—not a necessarily fatal flaw, but one worthy of consideration. Two additional strikes exist against silicon as the backbone of life. Carbon atoms are about ten times more abundant than silicon throughout the universe, so far as we can tell by examining the outer layers of stars and the material spread through interstellar space. Only with difficulty can we conceive of a situation in which life arises from silicon-based chemistry while a better alternative, an order of magnitude more abundant, lies at hand.

Furthermore, the cosmos contains not only carbon and silicon atoms but also large amounts of hydrogen and oxygen. The molecular bonds that carbon atoms form with hydrogen and oxygen have about the same strength as the bonds between carbon atoms, which permits the formation of complex molecules with multiple carbon-carbon bonds that provide the molecular backbone. In contrast, silicon atoms bond more strongly to both hydrogen and oxygen atoms than to themselves. Hence molecules based on silicon-silicon bonds are inherently unstable. Any straight-chain molecule with more than three silicon atoms linked together soon deforms, breaks apart, and forms other molecules. This contrasts markedly with the multiple linkages of carbon to carbon within straight- and branched-chain molecules, which dominate the chemistry of life on Earth.

Equally damning for silicon is the silicon atom's affinity for oxygen. Once silicon and oxygen atoms link together, they tend to remain bonded in small molecules, which can repeat themselves over and over to form stable, long-chain polymers and other structures but offer no possibility for the complexities of life. We live on a vast collection of these bonds: Most of Earth's rocks are silicates, silicon-oxygen compounds laced with other common atoms. Silicates, which remain chemically inert for geologically long intervals of time, are highly unlikely to undergo the complex chemical interactions that lead to living systems. Since oxygen atoms dominate silicon in abundance by a factor of twenty or so, essentially all the silicon on Earth became "locked" into silicate rocks soon after the Earth formed. There it will remain, and it seems likely to have undergone the same history on other planets with temperatures similar to Earth's.

The Need for a Liquid Solvent

So far as life's possible chemistry goes, carbon forms long-chain molecules, while silicon ends up in the small, tough molecular structures we familiarly call rock. Suppose that we therefore conclude that other forms of life are likely to use carbon as the crucial atom in forming their molecular structures. Must they also employ a liquid solvent, as life on Earth does?

Let us pause here for a paean to water. Every form of life on Earth essentially consists of one or more cells—sacs that contain water, which provides a solvent, a medium in which larger molecules float freely. (The root of the word *solvent*, like that of the word *loosen*, deals with dividing or setting loose.) As a solvent, water plays at least three crucial roles. First, it allows molecules to encounter one another, which they could hardly do if they were, for example, locked into the solid matrix of a quartz crystal. Second, the solvent protects against changes in the temperature which would otherwise kill an organism, especially a relatively complex one, by causing its molecules to break apart. The presence of a solvent buffers the effects of a temporary change in temperature, since the solvent as well as the crucial molecules must both be heated, rather than the molecules alone. Third, the solvent's surface tension—its tendency to form droplets—must have helped to organize life's processes before cells with complete membranes had evolved.

Of these three roles, the first must rank as the most important; life as we know it requires a liquid medium to carry the molecules whose interactions make life happen. Roles two and three, though important, might not prove crucial; other forms of life might, for example, exist in environments (like some on Earth) with extremely stable temperatures and might have formed through processes in which surface tension was not needed. If we allow ourselves to be convinced, however, that all life requires a liquid solvent, the natural question emerges, Why water?

Water, made of hydrogen and oxygen atoms, simply appears to be the finest solvent, not necessarily the only one. When we assess other potential solvents for life elsewhere, a reasonable restriction seems to arise from the relative abundances of the elements. Hydrogen chloride, for instance, might make a good solvent, but because chlorine ranks far down the list of cosmically abundant elements (not even in the top twenty, whereas hydrogen is number one and oxygen number three) a world where life relies on hydrogen chloride as its basic fluid seems unlikely. If we restrict ourselves to cosmically abundant liquid possibilities, we find that the six most abundant elements (hydrogen, helium, oxygen, carbon, neon, and nitrogen) can produce only three really likely liquids: water, ammonia (NH_3), and methyl alcohol (CH_3OH). Ammonia molecules each consist of three hydrogen atoms linked to a nitrogen atom, while molecules of methyl alcohol, a bit more complex, each contain one carbon atom, one oxygen atom, and four hydrogen atoms.

Like water, ammonia and methyl alcohol might prove to be

good solvents for living creatures, though water wins the head-to-head competitions. Of the three types, water molecules have the highest surface tension and furnish the best buffer against temperature changes, because it requires more energy to change the temperature of a gram of water than one of ammonia or methyl alcohol. Furthermore, water is likely to be significantly more abundant than ammonia or methyl alcohol on any Earth-like planet, because water probably forms the most common concentrate, the most abundant solid, in any planetary system. The processes that form planets do not allow liquids to exist until the planets have grown to considerable size. Hence a planet can produce rivers, seas, and lakes only by acquiring its future liquids in their solid or gaseous states, either as components of a flood of objects that bombard the newly formed planet or by "outgassing" them from hot layers below the planet's surface that contain the solvent in solid form.

Water has an additional advantage over competing solvents: Alone among all common liquids, water expands when it freezes! As a result, the ice we know floats on liquid water, but the ice formed from ammonia or methyl alcohol, denser than the liquid from which it freezes, would sink to the bottom of (hypothetical) lakes and oceans, which would therefore freeze from the bottom upward. Therefore, unlike the life in Earth's lakes and oceans, much of which can survive when the lakes freeze over or an icy layer forms on the sea, no possibility would exist for life to exist under the ice of other solvents. This could easily prove unimportant, however, since most life on Earth, even most life in the oceans, never has to worry about freezing.

Of course, any freezing of an ammonia ocean would occur only at temperatures below −78 C (−109 F), while a methyl-alcohol lake freezes only at −94 C (−147 F)! Not for nothing do we use methyl alcohol as an automobile antifreeze on Earth. However, the concept remains clear: If life relies on a particular liquid solvent, the solvent must remain liquid. For water, this requires temperatures between 0 and 100 C; for ammonia, between −78 and −33 C; for methyl alcohol, between −94 and 65 C. Here is one competition that water does not win, since methyl alcohol remains liquid over a wider temperature range than water does. But water has already won so many prizes that this difference barely handicaps it in the contest for best solvent.

Liquid Solvents and the Habitable Zone

In any case, the conclusions we can draw by assuming that life requires a solvent do not depend strongly on which of the three possible solvents life actually uses. The absolutely crucial point to draw from the analysis above turns on a single word: So long as living organisms use a solvent, they can flourish only within the temperature range where the solvent remains *liquid*. At temperatures outside this range, the solvent will either freeze or boil away, with fatal consequences. What does this mean in assessing planets' suitability for life?

In the days when we knew less, this line of thought led to the concept of the *habitable zone*, the region surrounding a star within which the heat from the star maintained a temperature that allowed life's solvent to remain liquid. For example, with water the solvent of choice, the sun's habitable zone was held to extend from somewhere inside the orbit of Venus to somewhere past the orbit of Mars (Color Plate 6). At distances too close to the sun, such as Mercury's, water could exist only in the form of water vapor, never as a liquid, and at distances too far from the sun, only ice, never liquid water, would appear. If we considered ammonia or methyl alcohol to be the key solvent, the inner and outer boundaries of the habitable zone would both move outward from the sun, since both of these compounds have freezing and boiling points that are lower than water's. Since the heat from the star declines rapidly as we move outward, the shifts in the boundaries of the habitable zone that arise from changing the solvent are relatively modest. To a reasonable approximation, we could state that the sun's habitable zone extends roughly from the orbit of Venus, at about 0.7 A.U. from the sun, outward past the orbit of Mars and almost to the asteroid belt at 2.8 A.U. All that remained would be first to calculate the habitable-zone boundaries for other stars, which would be larger or smaller than the sun's depending on the star's luminosity compared to the sun's, and then to see whether those stars had any planets orbiting within their habitable zones.

How Seriously Must We Take the Concept of the Habitable Zone?

Although the habitable-zone concept still makes sense, recent explorations of the solar system—and of our own planet!—make it clear that we must significantly readapt our mental picture to fit the more complex circumstances we have discovered. First of all, the sun's heat warms only one side of a planet at a time. This makes little difference on a rotating planet with a significant atmosphere, because the atmosphere convects heat from the day side to the night side. On Earth, the atmosphere means that the day-to-night differential equals only 20 to 40 F, not much when considering a possible range of many hundred degrees. But a planet with little or no atmosphere will show a much larger temperature differential, particularly if it rotates only slowly. The planet Mercury, for example, which has essentially no atmosphere, and on which the days and nights are each 88 days long, bakes at 600 F on its day side but freezes at −150 F on its night side. Close to the poles of Mercury, where the sun's rays arrive nearly horizontally, regions probably exist with Earth-like temperatures. Mercury itself almost certainly has no forms of life because it lacks an atmosphere, but the concept of polar life bears remembering when we assess the chances for life in other planetary systems.

A phenomenon far more important than the polar shadow for creating opportunities for life resides in the fact that *many planets have internal sources of heat.* On Earth, radioactive elements in the rocks slowly decay into other elements, releasing heat as they do so. This heating drives seafloor spreading (plate-tectonic motions), with the release of hot magma in volcanic eruptions all along the boundaries of the crustal plates, as well as at hot spots in the relatively thin crust beneath the oceans. Within the past two decades, undersea explorations have revealed deep-sea vents, from which enormous amounts of bacteria-rich hot water pours forth each second to join the much colder waters of the deep ocean. A wealth of bacteria have established themselves around these vents. Most of them belong to a newly discovered class, the Archaea, whose name reflects their apparent tremendous age. In fact, some biologists believe that life on Earth originated in deep-sea vents, completely independent of any solar heat and light! The Archaea have a biochemistry basically identical to other forms of life on Earth—for example, they use DNA in reproduction just as other organisms do—though their metabolism, as one would expect, does not use photosynthesis, which keeps many types of bacteria and plants alive. Instead, the Archaea "make a living" through chemical reactions that release energy—reactions that often involve methane molecules, which are released in high concentrations by the deep-sea vents. Other planets presumably also have radioactive rocks that release heat, so we can imagine that on some worlds, what would otherwise be frozen oceans remain liquid in their depths, kept warm, at least in part, not by stellar heating but by radioactivity.

Other possibilities also exist for heating a planet or its satellites. Io, the innermost large moon of Jupiter, heats through tidal flexing! Jupiter's enormous force of gravity produces a tidal bulge on Io, attracting the moon's near side with more force than the center, and the center with more force than the far side. If only Jupiter and Io existed, Io would bulge by a mile or two toward and away from the giant planet and would placidly move in a circular orbit around Jupiter, bulge and all, always keeping one side toward its master. But Jupiter's next two large satellites, Europa and Ganymede, also exert gravitational forces on Io, and these forces change as the relative positions of Io, Europa, and Ganymede vary during their orbits around the planet. The forces from

Europa and Ganymede keep Io from having a perfectly circular orbit: On each of its trips around Jupiter, Io moves a little farther from, and a bit closer toward, its planetary overlord. As this occurs, Jupiter's gravitational force on the tidal bulge changes, so the bulge—and the entire interior of Io—flexes and strains in response to these variations. Like the bending of a metal coat hanger, this flexure heats Io, so much so that hot jets of sulfur-rich material gush from numerous volcanic vents on its surface. Unlike Earth's volcanoes and the extinct volcanoes on Mars, which presumably also arose from the heat generated by radioactive elements, Io's volcanic activity comes from a complicated gravitational interplay among three different nearby objects. Complex though this interaction may seem, we can easily imagine that planets around other stars likewise have sets of satellites that demonstrate Io-like behavior as the result of their mutual gravitational forces.

Even more intriguing than Io is Europa, the next large satellite of Jupiter, where a crust of ice completely covers this miniature world about the size of our moon. This crust may cover a worldwide ocean, kept from freezing by the protection that the crust provides. Like Io, Europa's interior heats as the result of tidal flexing, and therefore has the potential to maintain liquid water beneath its crust. Should this hypothesis prove correct, Europa would rank as the likeliest site for extraterrestrial life in the solar system. The *Galileo* probe now in orbit around Jupiter has obtained high-resolution photographs of Europa that show changes in the long, thin cracks on the crust's outer surface—changes that could arise from the seepage of liquid water from below. But a new mission to the giant planet will be needed if we hope to obtain definitive information about what lies beneath Europa's frozen crust (Color Plate 19). Such a mission, which might land on the crust and drill beneath it, would determine whether Europa does have a liquid blanket and could begin to examine what processes have occurred during the more than four billion years of Europa's history.

Still more possibilities than those presented by Io and Europa could permit liquids to exist outside a star's conventionally defined habitable zone. The most important of these (based on our present knowledge) are atmospheric blankets that trap heat and warm their planets in a *greenhouse effect*. This effect arises from molecules in a planet's atmosphere that absorb infrared radiation emitted by the surface. Even though the infrared radiation temporarily trapped by the atmosphere does eventually escape into space, it delivers energy to the lower atmosphere before it does so. As a result, the greenhouse effect keeps the temperature on the surface and in the lower atmosphere higher than it would be in the absence of atmospheric molecules that absorb infrared radiation.

Among all the molecules that can be made from the most abundant elements, the most efficient greenhouse gases are carbon dioxide and water vapor. Mars, with an atmosphere less than 1/100th as significant as Earth's in its density and surface pressure, nevertheless has a noticeable greenhouse effect, because that atmosphere consists mainly of carbon dioxide. Venus, whose atmosphere likewise consists almost entirely of carbon dioxide, has a *600-degree greenhouse effect*, because the Venerean atmosphere has a density a hundred times that of our own and contains 10,000 times more carbon dioxide than Earth's.

On Venus, the greenhouse effect has turned what might otherwise have been a warm but perfectly pleasant planet (from the point of view of those who love liquid water) into a hellish nightmare, the hottest planetary surface in the solar system. The fact that Venus takes 244 days to rotate matters naught: The thick, haze-blanketed atmosphere, through which sunlight can never penetrate directly, spreads heat efficiently from the day to the night side and from pole to pole, keeping the entire planet baking at 750 F. On the other hand (for speculation never lacks an additional hand), if circumstances had placed Venus at, say, 2.5 to 5 times Earth's distance from the sun rather than its actual 0.72 times, and if Venus had developed the same atmosphere there that it now possesses, then Venus's greenhouse effect could allow water to exist as a liquid in regions where otherwise the temperature would remain forever far below freezing.

Radioactive rocks, polar shadows, tidal heating, the greenhouse effect. Does anything else complicate the habitable-zone concept? Certainly: the internal heating of a planet from its slow contraction. Planets made of rocks do not contract; their innards have sufficient strength to resist the gravitational forces with which every piece of the planet attracts every other piece. But a gaseous planet, held together by similar gravitational forces, does contract slowly from the effect of this self-gravitation, which generates heat. Jupiter, the sun's largest planet, has been undergoing such a contraction ever since it formed; so too, to a lesser degree, have the less massive gas giants Saturn, Uranus, and Neptune. All four of these planets have solid cores that do not shrink, but the bulk of their masses, surrounding the cores, consists of gas that indeed does contract. Jupiter, for instance, has a solid core deduced to contain about 10 or 15 times the mass of the Earth, leaving more than 300 of the planet's 318 Earth masses in the form of gaseous layers surrounding the core. On the other hand, these giant planets have no solid surfaces, which may provide a nearly total bar to the existence of life there (see page 38).

The *Galileo* Probe to Jupiter

Astronomers' models of Jupiter's internal structure imply that a dive into Jupiter would be a journey to frydom, a trip through ever hotter as well as ever denser layers of gas, penetrating to temperatures rising to many hundred degrees and pressures reaching thousands of times the atmospheric pressure on Earth's surface, writing *finis* to the history of the daring diver. As the *Galileo* spacecraft headed for orbit around Jupiter, it released a probe bound for just this sort of glory. On December 7, 1995, when the probe fell into Jupiter, encountering regions with progressively higher densities and pressures, it broadcast data on what it found for an hour or so, until the pressure rose above 20 atmospheres and the temperatures rose above 600 degrees Fahrenheit, rendering the transmitter inoperable. The probe's observations were captured by the orbiting spacecraft and slowly transmitted to Earth (the failure of the spacecraft's main antenna to open during the six-year trip to Jupiter greatly reduced the data-transmission rate, but all the essentials nevertheless came through), where they enriched our understanding of the first few hundred kilometers into Jupiter. Within half a day after its entry into Jupiter's atmosphere, all of the probe's components had been evaporated by high temperatures deep within the planet. However, the few hundred kilometers that the probe sampled include regions where the temperature equals what you feel now. A few tens of thousands of kilometers above, in near-Jupiter interplanetary space, the temperature holds constant at about −160 C (−258 F); a few tens of thousands of kilometers farther in, the temperature rises past 1,500 C (2,700 F); but in some intermediate regions, room temperature prevails.

At these temperatures, liquid water, ammonia, or methyl alcohol could exist—but they have nowhere to collect into ponds or lakes. Only if Jupiter offered *surfaces* on which sizable pools of liquid could form could the conditions be right for life to originate in much the same manner that life apparently arose on Earth, through countless interactions among molecules that produced an increasing variety of more complex molecules. But without surfaces, the most one could expect to find in Jupiter should be droplets of liquid. Astronomers rate the chances of finding liquid-bearing surfaces on or in Jupiter at just about zero, since the *Galileo* probe passed completely through the region where liquids could exist without encountering anything like a solid surface. This does not rule out all possibilities for life, but it does seem to make life a much more haphazard proposition, for if droplets were to combine into anything larger, they would tend to "rain" inward, falling into the death-dealing heat closer to the planet's core. But if life did somehow begin, it might have evolved parachute-like structures that allow even relatively large creatures to float through Jupiter's outer layers. Similar considerations apply to the other three giant planets—and to similar planets that orbit other stars.

Where Can We Expect to Find Pools of Liquid Solvents?

To answer the key question, Where do we find the liquids that can collect on a solid surface?, we have supplemented "a star's habitable zone" with at least five additional mechanisms—radioactive rocks, plate tectonics, the greenhouse effect, tidal flexing, and contraction of gas giants—that might permit the existence of liquid solvents in regions inside or outside the habitable zone defined solely on the basis of a planet's distance from its star. Notice, however, that heat from gas-giant contraction offers only a poor possibility, since gas-giant planets do not appear to have solid surfaces. Scientific speculation about extraterrestrial life is an arena of probabilities, so those who engage in it would certainly like to learn which is the most common source of heat for pools of liquids, and therefore quite probably for life, throughout the universe. Intriguingly, even the one form of life we know provides no easy answer for extrapolation. Most forms of life on Earth depend on solar heat, which favors the habitable-zone test, with only a modest increase from our atmospheric greenhouse effect. Some forms of life, however, including those that may be the original terrestrial life form, rely on geothermal heating, the heat released by radioactive rocks, and might have begun even if Earth lay outside the sun's habitable zone. We may yet find planets where solvents remain liquid, and where life draws its chemical energy, not from just one or two, but perhaps from three or four different possibilities.

Amid this panoply of doubt, some firm conclusions emerge. The habitable-zone analysis represents a conservative approach, since if we use only this test, we can never overestimate the possibilities for regions where life might exist. Hence when we find a planet that orbits within its star's habitable zone, we have struck gold (metaphorically speaking), or at least an auriferous layer, in our search. But when we find a planet outside the habitable zone, we should not immediately conclude that it must be lifeless. Such a planet might use one of the five possibilities described above (or others that we have yet to encounter) to maintain a liquid solvent.

Ancient Life on Mars?

The conclusions we have reached from a habitable-zone assessment of the possibility of life finds some support in the recent evidence for ancient life on Mars. During the 1980s scientists realized that some of the meteorites found on Earth's surface actually originated on Mars, from where they were blasted into orbits around the sun by titanic impacts on the red planet. If our planet's orbit happened to cross the path of this debris, some of it may fall to Earth—pieces of Mars readily available to those who can find them! Mars meteorites can be identified through careful chemical analysis, which reveals that they contain gas trapped in tiny pockets whose composition corresponds exactly to the composition that the *Viking* spacecraft measured in the Martian atmosphere but differs markedly from our own. These measurements, which include relative abundances of the different isotopes of inert gases such as argon and xenon, provide a forensic "smoking gun," capable of convincing a jury that a small fraction of all meteorites do have a Martian origin.

In 1994 a team of geologists and biologists began a detailed study of one of the Mars meteorites, cleaving it into superthin slices and subjecting it to chemical analysis and laser-electron microscopy. Two years later, in August 1996, they announced their startling discovery: This Mars meteorite, a four-pound rock named Allan Hills 84001 after the Antarctic mountains where it was found, speaks of possible ancient life on Mars! The fourfold evidence for life consists of carbonate-laced globules (carbon-oxygen mixtures associated with life on Earth); two types of minerals, magnetite and iron sulfide, which bacteria often produce as part of their metabolic processes; a particular variety of carbon-based molecule, called polycyclic aromatic hydrocarbons (PAHs), which small organisms on Earth produce, though these molecules also arise from nonbiological processes; and tubular structures less than one micron across that may be microfossils, actual structures of long-vanished creatures on Mars. Each piece of evidence testifies in silent eloquence; collectively, they speak volumes to the scientists who can read their messages and who have judged this evidence as compelling, though not conclusive. The debate over the rock from Mars will continue, with skeptics pointing out that the more prosaic, nonbiological explanation must be favored until it can be definitively ruled out. But the thought of ancient life on Mars remains provocative.

Radioactive dating of the Mars meteorite points to an age of three to four billion years since the globules within the rock formed—three-quarters of the way back to the formation of the sun and its planets. Any tiny Martian organisms in Allan Hills 84001 lived during the epoch when life on Earth still consisted solely of microscopic organisms. In those years, not long after the solar system formed, the rocks blasted from a planet by repeated large impacts might have carried terrestrial life to Mars, or Martian life to Earth. "Who is to say we are not all Martians?" has become a good sound bite to promote the news from Mars.

However, even if later research verifies the discovery of ancient life on Mars, we must recognize that today Mars offers far less hospitality to life than the planet did four billion years ago. During the first two or three billion years after the solar system formed, while life on Earth flourished only in the primitive oceans, Mars, which then as now was far less massive than Earth, gradually lost most of its atmosphere. This loss prevented Mars from maintaining liquid water on its surface, because the long-term existence of any substance in liquid form requires a minimum atmospheric pressure, below which any liquid promptly evaporates. On Earth's surface, the pressure falls below the point that would allow carbon dioxide (CO_2) to exist as a liquid, so we can watch solid CO_2 (dry ice) sublimate, as chemists say, directly from the solid into the gaseous state. On the surface of Mars, where the atmospheric pressure equals only 0.7 percent of its value on the Earth's surface, water would behave the way that carbon dioxide does on Earth.

From photographs taken by the *Viking* orbiters in 1976, which reveal sinuous channels apparently carved by running water, we know that Mars had rivers (and presumably also lakes or seas) throughout at least its first billion years. As liquid water disappeared from the Martian surface, one of two scenarios therefore must have played itself out: Either life vanished completely, or else some forms of life managed to survive by colonizing specialized environments, perhaps in permafrost below the exposed surface or at the edges of the polar caps.

If the fossil and chemical record within Allan Hills 84001 should be confirmed as a sign of life, biology will enter a new era, in which we can contrast the forms of life that developed in two astronomically distinct locations. Just as one extrasolar planet might be an anomaly, so too one single form of life—life on Earth—might prove a singleton. If scientists can establish the existence of life, even long-vanished life, on our neighbor planet, the similarities and differences between the Martian and terrestrial forms of life should tell us how easily, and thus how often, life may have arisen throughout our solar system and beyond. As chance would have it, at the time when scientists announced the results of their analysis of Allan Hills 84001, NASA had two spacecraft almost ready to embark on journeys to Mars, to arrive during the second half of 1997. One of these, the *Mars Global Surveyor*, will map the entire planet in finer detail than ever before. The other, called *Mars Pathfinder*, will carry a remote-controlled rover that can roam the Martian surface, searching for rocks that may provide important information about microbes that even now may exist on Mars. A Russian spacecraft, *Mars 96*, unfortunately failed in its mission upon launch in November 1996.

Must Life Be Restricted to Planets?

So far, relying on the temperatures and molecular mixtures that planets provide, we have dealt with planets and their large moons as the only possible locales for life. Shouldn't we ease this restriction and allow the possibility of life in interplanetary space, or even life that sprawls through the vast spaces between the stars, carrying stored supplies of energy? (See Fred Hoyle's *The Black Cloud* for an imaginative science-fiction exposition of the second possibility.) At least two reasons exist for confining ourselves to planets, at least for the time being. First, we certainly best understand life on a planet. Second, and more important, much can be said in support of two requirements for life to originate: relatively large amounts of liquid; and surfaces on which molecules can collect. The call for surfaces comes from theories of how life began, not simply through molecular interactions in midpuddle, but rather at places where the liquid meets the edge. There, on the shoreline, molecules can concentrate through repeated cycles of wetting and drying, and there too, the structure of the solid surface can impart order to the liquid-borne molecules that periodically collect as the shoreline advances and recedes.

The need for large amounts of liquid has a simple, probabilistic basis: Small amounts offer fewer possibilities for random interactions to produce complex molecules and, eventually, the replicating and evolving systems we call life. However, to maintain liquid in large amounts (pool-sized, for instance) requires large objects. Any liquid evaporates, and if it does so on an object that exerts only modest gravitational forces, it will escape completely. Only a large object can retain this evaporating material in its atmosphere, with some of it, at any time, raining droplets back onto the surface for later evaporation.

Two factors, temperature and mass, govern whether or not a planet can maintain a relatively thick atmosphere through billions of years. At low temperatures, even an object not much larger than our moon can retain a thick atmosphere, as Titan does in the Saturnian system. At high temperatures, similar to those at Mercury's distance from the sun, even the Earth, with a mass 40 times Titan's and 81 times the moon's, would lose its atmosphere in a few billion years' time. On the other hand, a planet with a mass similar to Jupiter's, like the planets around 51 Pegasi, Tau Bootis, and Upsilon Andromedae, could retain an atmosphere even at only one-tenth of Mercury's distance from the sun.

We may conclude that keeping a solvent in liquid form requires objects at least as large and massive as our moon—even more massive if the object orbits at an Earth-like distance from a sunlike star. We must not forget, however, that exceptional circumstances such as those that may exist on Europa can provide pressure without having a thick atmosphere. In our solar system, we find more than a dozen objects that pass the mass test. However, if we impose the additional requirement that the object must have a *surface*, we eliminate the four gas-giant planets. Then, since Mercury is too small to hold an atmosphere at its distance from the sun, and since Io, Ganymede, and Callisto have no real atmospheres either (Io does have some atoms spewn from volcanoes surrounding it at any time, but these quickly escape into space), the number of objects with solid surfaces and potential liquid solvents falls to five: Venus, Earth, Mars, Titan, and Europa, which we include on the supposition that its icy crust conceals a worldwide ocean and prevents it from evaporating. Of these five "good" objects, one planet (Venus) has no liquid, thanks to its stifling blanket of atmosphere, and another (Mars) has such a thin atmosphere that no liquid can exist on its surface. Both these "failures" represent close cases. Venus has experienced a runaway greenhouse effect that, astronomers calculate, might well have been avoided if the planet had a distance from the sun 15 to 20 percent greater than its actual distance. On Mars, channels carved by running water billions of years ago testify to the fat that the red planet once had conditions allowing a liquid solvent, but lost them. If Mars had twice its actual mass—1/5 of the Earth's mass rather than 1/11—it might still possess clouds laden with water vapor, rainstorms, and pools of liquid water.

Life on Titan?

Even if we discard Venus and Mars, and even if Europa should prove to have no liquid beneath its icy crust, we may conclude that our solar system offers more than a single site with a possible liquid solvent. Saturn's large moon, Titan, reminds us that life could exist on an object much colder than Earth, not a planet but a planetary satellite. Titan and Saturn orbit the sun at almost 10 times the Earth-sun distance, so far from our star that the sun's apparent brightness falls by a factor of 100. In addition, Titan's atmosphere filters out most of the sun's diminished glow, because it contains long-chain polymer molecules that create a red-brown haze analogous to the smog above the cities of Earth. Because Titan's smog-laden clouds permanently veil its surface from our view, if we hope to learn much about its surface, we must construct, launch, and guide an automated spacecraft that can pierce Titan's opaque veils with radar and, even better, automated landers. Humanity has put the second option into operation: NASA (as the junior partner) and ESA, the European Space Agency, have constructed a probe to Saturn called *Cassini-Huygens*. *Cassini-Huygens* has a scheduled launch date in late 1997 and should, if all goes well, tell us about Titan's surface in the year 2002.

We already know that Titan resembles Earth in having a thick, nitrogen-rich atmosphere; Pluto and Triton, Neptune's large satellites, have mostly nitrogen atmospheres, but these are amazingly tenuous. Contrary to Kurt Vonnegut's delightful fantasy *The Sirens of Titan*, the atmosphere of Saturn's largest satellite contains no detectable oxygen (see chapter 9 for the negative implications that this imposes on the search for familiar types of life). Titan's atmosphere does have some argon and methane, as well as trace amounts of hydrocarbon molecules (carbon-hydrogen mixtures), carbon monoxide and carbon dioxide, and some molecules made of carbon, hydrogen, and nitrogen, such as hydrogen cyanide (HCN) and cyanoacetylene (HC_3N). Ten times farther than Earth from the sun, Titan has a surface temperature of about −180 C (−356 F), far colder than any temperature recorded on Earth. Yet Titan offers the best solar-system possibility for finding pools of liquid on its surface. This occurs because of Titan's thick atmosphere, which provides sufficiently high pressures for liquids to exist.

What substances remain liquid at 356 below? The most likely molecules to provide this liquid are ethane (C_2H_6) and propane (C_3H_8), hydrocarbons that are known to exist in Titan's atmosphere. Note that we created our list of the "big three potential solvents"—water, ammonia, and methyl alcohol—on the supposition that life would most likely originate at temperatures between −150 and +100 C; the fact that hydrocarbons such as ethane and propane did not make the list reminds us that we always risk error by assuming that a few possibilities cover all the options. Astronomers speculate that Titan's surface most likely consists of thoroughly frozen ice (H_2O), crossed by rivulets of ethane, propane, and other hydrocarbons, running downhill across the ice into pools, from which the hydrocarbons evaporate—even at these low temperatures—eventually to return after completing a "rainfall" cycle similar to Earth's. If astronomers' current picture of Titan proves correct, then even though chemical interactions occur far more slowly in a hydrocarbon pool at −356 F than within a tide pool at +70 F, we cannot exclude the possibility that some form of life, totally different from life on Earth, might have had a chance to begin and to evolve on this distant satellite of a distant planet.

What Does This Tell Us about the Habitability of the Newly Discovered Planets?

Our survey of the solar system has revealed several extraterrestrial sites that seem at least marginally fit for life, a reminder that we could hardly expect our Earth to be the only suitable location. What can we say about the new planets discovered in orbit around sunlike stars? Suppose that all of our generalized conclusions prove valid (a situation not likely to occur in the real world). We then expect life on Titan, and on planets around other stars, to consist of long-chain molecules based on carbon, to rely on a liquid solvent, and to evolve into different forms through the billions of years that typify the age of a planet around a sunlike star. Although the habitable zones around such stars should offer the most likely regions to find life-bearing planets, we have noted that one of the five exceptions to the rule of the habitable zone (or another exception not yet considered) could permit life to exist outside the basic habitable zone. Notice, however, that we do not expect to find life *inside* the inner boundary of the habitable zone, because planets do not seem to have ways to cool themselves below the temperature that arises from the balance between the heat that the planet receives from its star and the heat that it radiates into space.

We should therefore imagine the habitable zone around a sunlike star as a band with a definite inner edge but only a vague outer boundary. The inner boundary lies about 0.7 A.U. from a star with the sun's luminosity, while the outer edge begins to fuzz out past about 1.5 A.U., with possible "islands" at any distance—provided that sufficient heat exists, independent of anything received from the star, to maintain solvents in liquid form.

Despite all this fuzzy talk, the verdict on the first extrasolar planets to be discovered is simple: Only one or two of them have the proper temperatures to support life, and these planets almost certainly are gas giants without surfaces on which liquids could collect and dry out. In temperature terms, only the planet around 47 Ursae Majoris, orbiting at 2.1 A.U. from its star, and the planet around 16 Cygni B, orbiting at 2.2 A.U., seem promising (and possibly the planet around 70 Virginis as well). However, if these planets resemble Jupiter, we encounter exactly the same problem that Jupiter presents: lack of a solid surface on which pools of liquid can form.

If astronomers allowed themselves to yearn, they would long for a few planets with clear signs of life (see chapter 9 for what those signs might be). They would then know, for example, whether the habitable zone should be taught in elementary schools, or whether it represents another fine concept shot down by additional knowledge. Learning rather than yearning must assuage the astronomers' psyches, and quite fortunately for them and for the rest of us, this learning may soon occur. We may confidently await the discovery of more planets by the Doppler-shift method, but these are likely to have Jupiter- and Saturn-like masses, and thus to be giant planets as well. To find planets with roughly Earth-like masses, we must employ other search techniques, described in chapter 6. And of all the methods for finding extrasolar planets, the one probably least likely to work during the near future is the most straightforward: using our finest telescopes to *see* planets. To understand why this won't work with our present abilities requires a chapter, conveniently placed immediately after this one.

CHAPTER

3

Why Can't We See *Planets?*

Of all the questions raised by the discovery of planets around nearby sunlike stars, the most obvious, the most touching, and the most plaintive (especially to astronomers who understand how public interest drives government funding) are these: Why can't we *see* the newly discovered planets? And why can't the world's finest telescopes, including the now-refurbished Hubble Space Telescope, search directly for more planets around other stars?

The answers to these questions lead us to constraints imposed on our vision by atmospheric blurring, the sizes of telescopes in space, the fundamental laws of physics, and limited funding. These constraints are worth examining in order to comprehend the majestically enormous task in which we engage when we decide to search for extrasolar planets. The greatest part of the search burden arises from three basic facts about planets orbiting other stars. Planets are small; they produce no light on their own but only reflect light; and they nestle so close to their stars that they are easily lost in the far greater glow of their stellar overlords.

Enormous Distances and Enormous Distance *Ratios*

We can easily accept the first of these facts as obvious—so long as we consider other worlds than our own. (See chapter 8 for a discussion of our inability to suppress our intuitive belief that Earth forms the center of the universe, with all else a modest canopy.) Matter in the universe has aggregated itself into galaxies, which themselves form clusters with thousands of member galaxies, as well as even larger aggregates such as the "Great Wall of Galaxies" that stretches for hundreds of millions of light years across the Northern Celestial Hemisphere. Large galaxies such as our own Milky Way have diameters of many tens of thousands of light years and contain hundreds of billions of stars. As a result, the average distance between stars in a typical large galaxy equals approximately one *light year*—the distance that light travels in a year, about 10 trillion kilometers, or 6 trillion miles. Since we inhabit the outer suburbs of the Milky Way and orbit a single star rather than a double-star system, we should not be surprised to find that the closest stars to the sun lie 4.4 light years away—but at that distance we find three of them.

Typical stars have diameters of approximately 1 million kilometers (for the sun, this value equals 1.4 million kilometers, or 800,000 miles), so *the average distance between neighboring stars equals about 10 million times the diameter of a star!* This large number means that if we were to model stars with light bulbs, each a few inches across, we must place the light bulbs in separate cities, separated by many hundred miles, so that the United States might contain a few dozen of these stellar bulbs. Of all the astounding facts astronomy offers to shape our perspective, none can surpass the simple truth about the enormous distances between stars. So far from star to star; so nearly empty are the vast interstellar spaces! Intuition cannot accept these enormous realms, nearly void of interest to the casual observer, though the question of how stars ever formed from the diffuse gas

Great Square of Pegasus

51 Pegasi

Planet around 51 Pegasi

Planet orbits at 0.05 A.U.
Orbit is circular (e = 0.0)
Minimum mass is 0.47 Jupiter masses
Orbital period is 4.23 days
Star is 45 light years from solar system

size of Earth's orbit

COLOR PLATE 1 The first extrasolar planet discovered around a sunlike star orbits the star 51 Pegasi, which lies close to the "Great Square" of the constellation Pegasus and is about 45 light years from the solar system. This planet takes 4.2 days for each orbit, in which its distance from 51 Pegasi is only 5 percent of the Earth-sun distance. The planet's mass equals at least 47 percent of the mass of Jupiter.

COLOR PLATE 2 The second extrasolar planet orbits 70 Virginis, a sun-like star about 72 light years from the sun that lies between Spica, the brightest star in Virgo, and Arcturus, the brightest star in the constellation Boötes. This planet takes four months to complete its noticeably elongated orbit, in which its average distance from 70 Virginis equals 43 percent of the distance from the Earth to the sun. The planet's mass equals at least 6.6 times the mass of Jupiter, and this object may yet prove to be a brown dwarf.

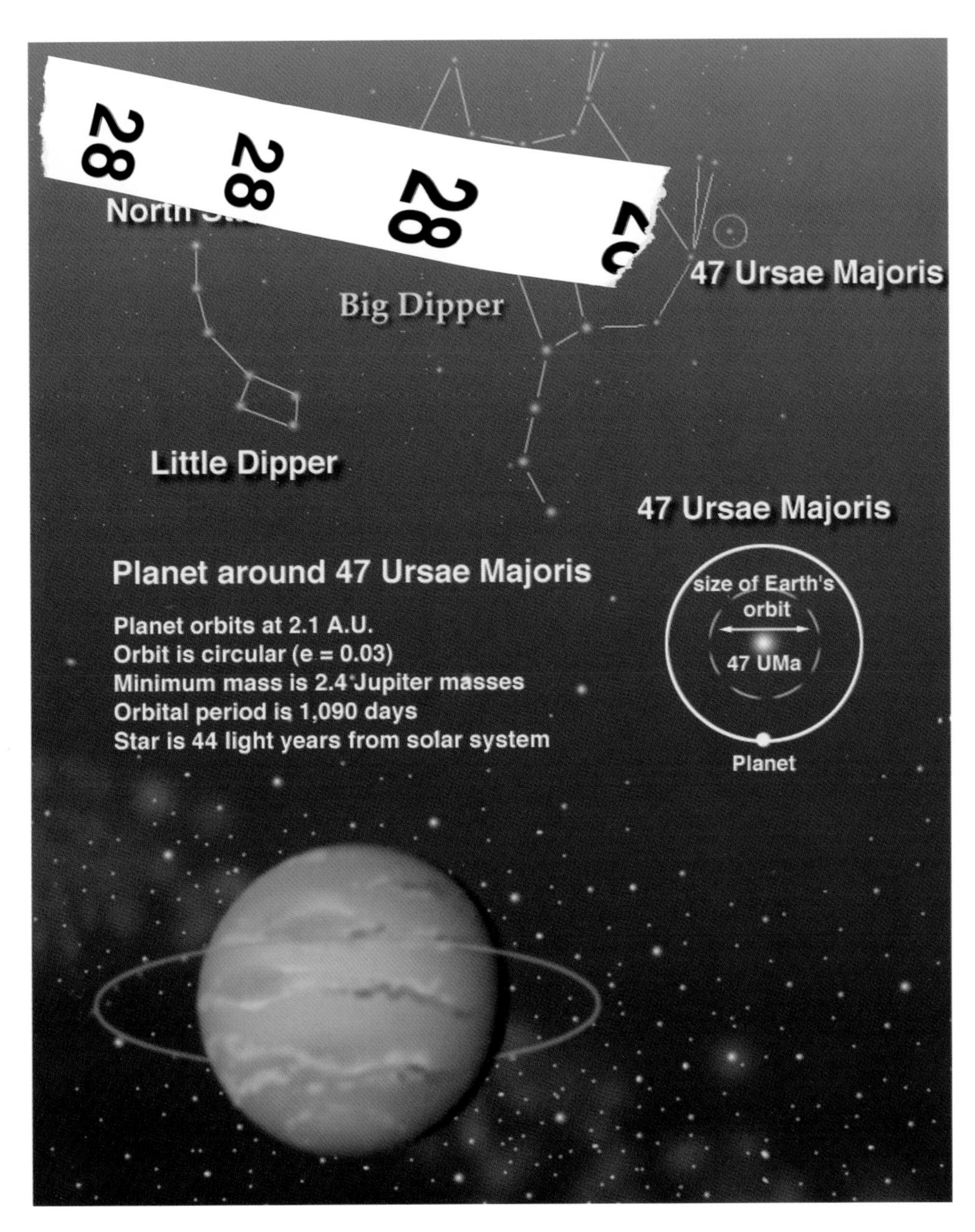

COLOR PLATE 3 The third extrasolar planet, around 47 Ursae Majoris, orbits once every three years around a sunlike star close to the Big Dipper, 44 light years away. The distance from the planet to its star is just over twice the Earth's distance from the sun, and the planet has a mass at least 2.4 times the mass of Jupiter.

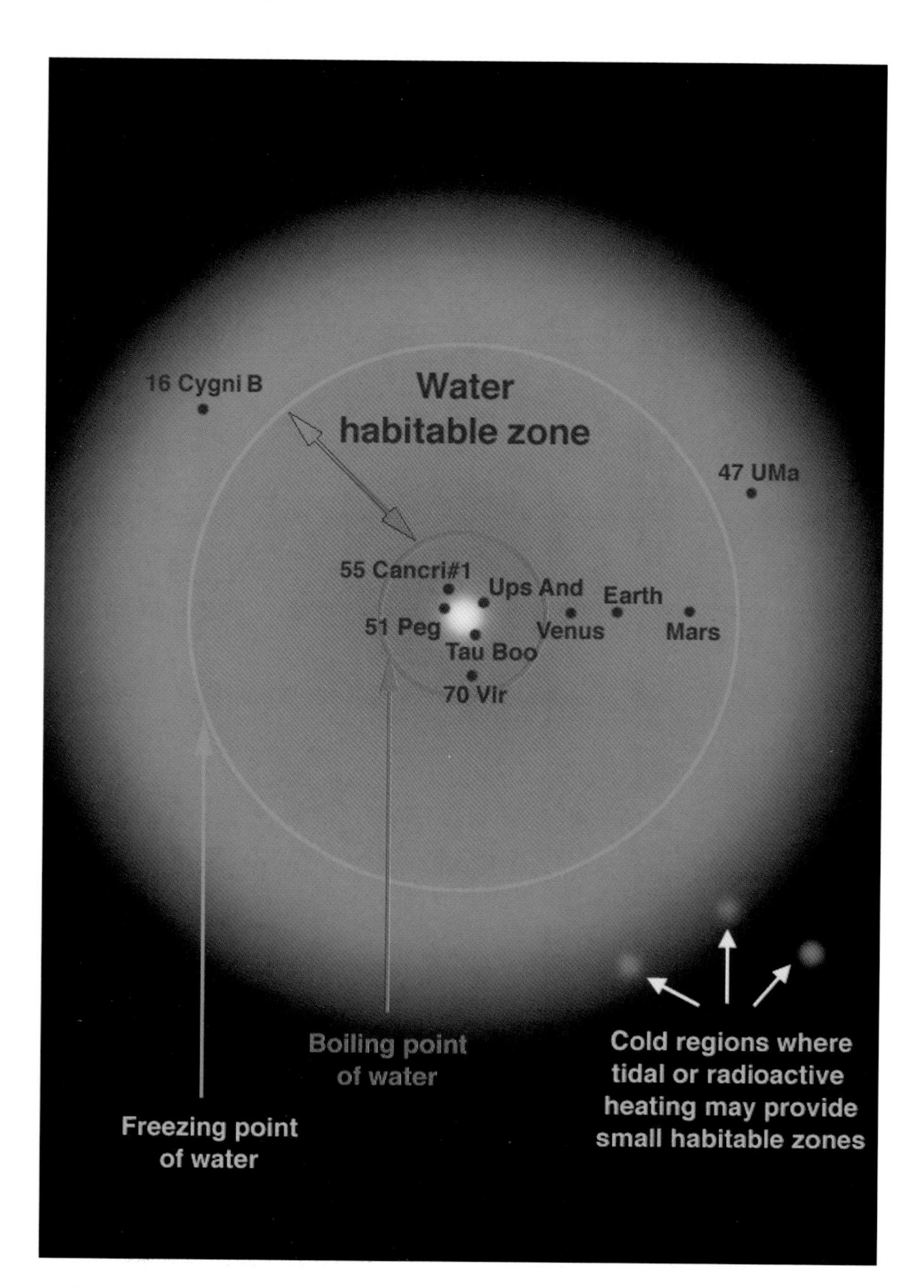

COLOR PLATE 6 The habitable zone surrounding a star includes the regions where the star's heat can maintain a liquid solvent such as water, ammonia, or methyl alcohol. In addition, local conditions may also allow a liquid solvent to exist outside the habitable zone. This diagram shows the inner and outer boundaries of the habitable zone around a sunlike star, assuming that water is the solvent for living organisms. The diagram also indicates the distances of the newly discovered planets from the stars that they orbit, using the star's names to designate the planets.

COLOR PLATE 7 In December 1995 the Hubble Space Telescope used 100 hours of observing time to secure the "deepest" image to date of the universe. This image, which demonstrates the high-resolution capability of the telescope's refurbished optical system, reveals the faintest objects ever glimpsed by humanity. Most of these objects are galaxies 5 to 10 billion light years away, seen as they were when the universe was much younger than it is now.

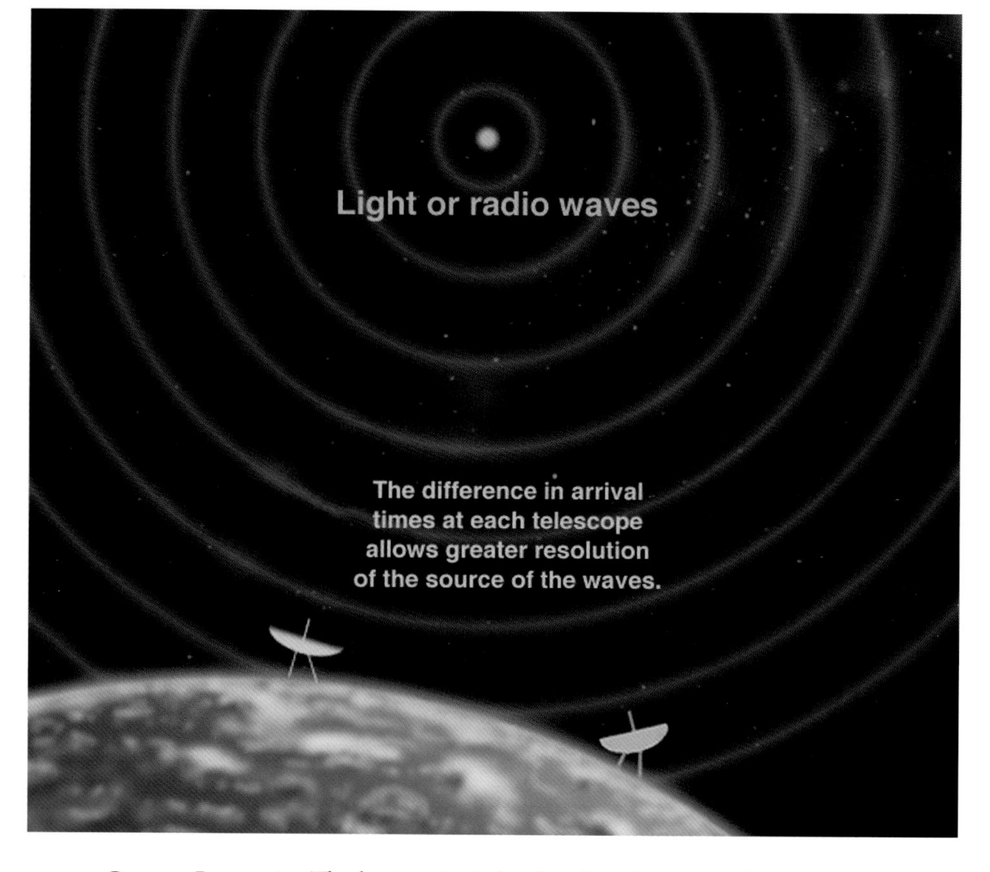

Color Plate 8 The basic principle of an interferometer is to use two or more detectors to observe a source of electromagnetic radiation (for example, of light or radio waves) and then to combine the observations made with each detector. With the correct treatment of the "interference"—the comparison of the signals from each detector—an interferometer can produce an image of the source that has the same angular resolution as an image produced by a single large detector whose diameter equals the maximum separation of the individual detectors in the interferometer.

spread throughout these regions fascinates astronomers. Accept them we must, however, if we hope to comprehend the cosmos.

The enormous sizes of interstellar distances completely dwarf those within the solar system. In fact, the word *dwarf* barely begins to describe the situation. For example, the average distance from the Earth to the sun, the astronomical unit or A.U., amounts to nearly 93 million miles—much too large to correspond to anything in our daily experience, but small enough to be imagined by extrapolation. If we drive a car at 93 miles an hour, we can travel 93 million miles in a million hours, a bit over 114 years. But when we speak of *interstellar* rather than interplanetary distances, we lose most of the benefit of these homely analogies. To travel 1 light year at 93 miles per hour requires not 1 million but 70 billion hours—not 114 but *8 million* years. Since we have no real feeling for a million years, the example begins to lose its force: All large numbers feel about the same to most people. This fact lies at the heart of an astronomical tragedy of public understanding, the failure to perceive that although all astronomical distances are enormous, some of them are so much more enormous than others that the *ratio* of distances itself demands separate, special consideration.

Consider what this means when we search for planets around other stars. Alpha Centauri, the closest star system to our own solar system, basically consists of two sunlike stars, Alpha Centauri A and B, orbiting their common center of mass and separated by 23 times the Earth-sun distance. A much dimmer, less massive star called Proxima Centauri, a faint red dwarf star, also orbits this center of mass, at a distance of several thousand A.U. The distance from the solar system to Alpha Centauri equals about 14,000 times the distance between Alpha Centauri A and B. At this distance and with this separation of its bright stars, even a modest telescope will reveal the separate glows of Alpha Centauri A and B despite the fact that the Alpha Centauri system lies 4.4 light years from the solar system. In angular measurement, which describes what we see on the sky, the two stars are separated by 18 seconds of arc. Sixty seconds of arc make 1 minute of arc, and 60 minutes make 1 degree. An angular separation of 18 seconds of arc amounts to 1 percent of the angular diameter of the sun and the moon, each of which equals 30 minutes (1,800 seconds) of arc.

Splitting Double Stars with a Telescope

We can use Alpha Centauri A and B as a benchmark with which to rate the difficulty of observing any planets that orbit other stars, all of which have greater distances from the solar system than the Alpha Centauri system. *Angular separation*—the angle between objects as we see them—follows a simple rule: The apparent angle separating any two objects decreases in direct proportion to the distance from the objects to the observer. Hence if the Alpha Centauri system had ten times its actual distance and lay at 44 rather than 4.4 light years from Earth, its two stars would have an angular separation of 1.8 seconds of arc rather than the actual value of 18.

The *angular resolution* of any optical system specifies the smallest angular separation of two points of light that can be seen as individuals, rather than as a single, relatively blurred point. Any reasonable amateur telescope can resolve—see as separate points of light—two stars whose angular separation equals at least 2 seconds of arc. This makes the actual Alpha Centauri system an easy target for angular resolution, but if the stars had a distance ten times greater, an amateur would have to invest in a better instrument to achieve this goal. Observing on a clear, still night, such a telescope would resolve the two stars as individuals. At a distance three times greater still (132 light years), the angular separation of the two stars would become 0.6 second of arc. Then even our finest telescopes would have great difficulty in seeing the double-star system as anything but a single point of light. At significantly greater distances, no chance would exist for Earth-based astronomers to see two separate stars in a system like Alpha Centauri.

Why not? Here two effects come into play. The Earth's atmosphere is never perfectly still; instead, its continuous motions refract or bend the light rays passing through, and by amounts that change each fraction of a second. This produces twinkling, an effect admired in many quarters but not much by astronomers, who know that the atmosphere stands between them and better angular resolution. Astronomers refer not to twinkling but to *seeing*, a sweet word that describes the best angular resolution attainable on a given night in a particular location. To diminish the role of atmospheric refraction and to improve the seeing, astronomers locate their observatories at high altitudes, strategically situated to place them above a significant fraction of the total atmosphere. The Mauna Kea Observatory on the Big Island of Hawaii, nearly 14,000 feet above sea level, has become the world's premier observing site largely because it stands above nearly half the mass of gas that forms our atmosphere. This causes problems for astronomers, who must acclimatize themselves each afternoon before observing and occasionally take a few puffs of oxygen from the bottles stored by every telescope, but it allows them, on fine, clear nights, to obtain *sub-arc-second seeing*, times when the atmospheric blurring diminishes to the point that astronomers can see as individuals two points of light separated by significantly less than one second of arc. Under such conditions, Mauna Kea's telescopes could reveal Alpha Centauri A and B even at distances out to about 200 light years. Notice, however, that all the money in the world, invested in ever-larger telescopes, cannot directly overcome the limitations imposed by seeing: Even the best telescopes remain subject to the continuously changing refraction of light rays traveling through Earth's atmosphere.

The Hubble Space Telescope and the Laws of Optics

Instead of building larger telescopes, we can invest in telescopes that entirely avoid the problem of seeing: telescopes sent beyond the atmosphere. Just this task fell to the Hubble Space Telescope (HST), conceived during the 1960s, constructed during the 1970s, in storage during the 1980s, sent into orbit in 1990, four years after the *Challenger* disaster had delayed all United States space launches, fitted with correcting optics at the end of 1993, and functioning brilliantly as our greatest single astronomical instrument since that time. The HST experiences essentially no effects from our life-maintaining blanket of air, since its orbit at an altitude of 300 kilometers keeps it above all but the outermost vestiges of our atmosphere. What, then, limits the telescope's ability to see double stars as individuals?

Only the fundamental laws of optics, from which there is no escape. Since light consists of electromagnetic waves, the task of distinguishing the waves from two sources close to one another on the sky consists of separating two streams of waves arriving from nearly identical directions. To do so, we need one instrument that collects the wave streams, another that detects them, and a third that analyzes them. In our own eyes, the lens collects light waves and focuses them onto the retina, which detects them and sends signals via the optic nerves to our brains, which analyze these detections. Astronomical telescopes use lenses or mirrors to collect and focus light, along with photographic plates, or CCD detectors, to record what the telescope collects, and a computer for analysis of the signals detected. CCD detectors are solid-state devices, now at the core of almost every video camera, that can detect and record light waves far more efficiently than even the finest photographic film; as a result, astronomers have almost universally turned to CCDs for detecting light waves. CCDs offer the additional advantage that they produce digitized signals, ready for computer processing, whereas photographic films and plates must be carefully "scanned" by photometric systems that measure the darkness of each part of the image.

The heart of any telescopic system remains the part that collects light—the primary lens or mirror, but almost always a mirror, because mirrors, which can be supported and adjusted from behind, can be made far larger than lenses, which must be supported around their edges. To discriminate between two light sources in close proximity, the telescope must collect light from as wide an area as possible. We can understand the situation by recognizing that each part of the lens or mirror has a slightly different perspective in its view. As these perspectives become more spread out, the chances improve of seeing two objects as individual sources rather than as a combination. You can experience this fact by inspecting the room in which you are sitting with only one eye and comparing this view with what you see with two eyes: Your ability to gauge distances decreases. But this distance-gauging ability, though highly useful, actually provides only a small part of the ability we would have if our brains had evolved to use the full capacity offered by two separate views of the world. With "perfect" evolution, we could use our binocular vision to see the world with a sharpness—that is, an angular resolution—far better than what we have now.

Science speaks in mathematical form. Let us therefore take a moment to admire the simple rules that govern angular resolution. For any system that collects electromagnetic waves, such as light waves, the theoretical best angular resolution equals approximately 250,000 seconds of arc times the ratio of the wavelength of the waves under observation to the diameter of the collecting system. The wavelength-to-diameter ratio implies that we can achieve better angular resolution in one of two ways: *either by decreasing the wavelength of the radiation under observation or by increasing the diameter of the light collector*. If we study the universe in short-wavelength radiation such as ultraviolet or x rays (an important aspect of modern astrophysics, but not strictly relevant here), we can expect an improvement in angular resolution solely from our choice of radiation. If, however, we observe in visible light, then the sole means of improving our angular resolution lies in building larger telescopes.

What angular resolution can we hope to achieve? Visible light has wavelengths that range from about 40 to 80 millionths of a centimeter; our eyes best detect yellow light with wavelengths close to 60 millionths of a centimeter. Because the lenses in our eyes have diameters of approximately 0.25 centimeter, the ratio of the wavelength of yellow light to the span of our light-collecting systems amounts to about ($60 \times 10^{-6}/0.25$), or 2.4×10^{-4}. If we multiply this number by 250,000, we obtain a theoretical best angular resolution for the unaided human eye equal to 60 seconds of arc. These 60 arc-seconds amount to 1 minute of arc, the angular size of a 3-foot-wide suspension bridge cable seen from a distance of ten miles. You can check this result in the vicinity of one of the world's great bridges, noting that you can see the gracefully bowed main cables as individual strands, but not the thinner, vertical cables that suspend the bridge deck from the main cables. The criterion we have specified—an angular resolution of 250,000 seconds of arc times the ratio of the wavelength to the collector diameter—technically refers to the ability to separate two equally bright sources of light, but it also describes rather well whether or not we can perceive a thin line, such as a faraway bridge cable.

Now suppose that we purchase a telescope with a lens or mirror 15 centimeters (6 inches) in diameter. If the telescope's optics have been carefully constructed, we would expect to attain an angular resolution 60 times better than that of our unaided eyes, not 60 seconds, but 1 second of arc, provided that the atmosphere cooperates by remaining clear and still. If we pass from amateur telescopes into the realm of professional instruments, we can quickly calculate that a 3-meter telescope (one with a 300-centimeter-wide lens or mirror) has a theoretical angular resolution of 0.05 second of arc, and that the giant Keck Telescopes, with 10-meter mirrors, can each theoretically achieve angular resolutions of 0.014 second of arc.

Telescopes as Light Collectors and Angular Resolvers

We might pause here to ask whether the astronomers have gone slightly haywire. The angular-resolution rule described above shows that a telescope with a lens or mirror 15 centimeters wide has an angular resolution of 1 second of arc, which is about the best we can hope the atmosphere to allow. Why, then, have astronomers built far larger telescopes, with 2-meter, 4-meter, 5-meter, and now the 10-meter-wide compound mirrors in the twin Kecks on Mauna Kea? Aren't astronomers wasting their efforts by constructing telescopes theoretically capable of an angular resolution better than 0.1 second of arc, since the Earth's atmosphere, even above a high Hawaiian mountain, only rarely offers "seeing" as fine as 0.3 second of arc?

The answer to this objection lies in the fact that telescopes perform more functions than providing angular resolution. They also serve as "light buckets," collectors of light from faint sources. This light-gathering function allows large telescopes to detect fainter sources than small ones can; in a small telescope, the light that arrives at the collector is insufficient to permit the detector to distinguish the source from the background of stray light, which always exists, because the night sky glows with diffuse light, and the telescope itself scatters some light into the field of view. Light-gathering power, far more than improved angular resolution, provides the impetus behind the construction of Earth's largest telescopes. Because of the blurring effect of our atmosphere, the 10-meter-wide mirrors of the Keck Telescopes may not obtain an image of a particular object any crisper than the image from the nearby Canada-France-Hawaii Telescope, with a mirror "only" 3.8 meters wide. On the other hand, either of the Keck Telescopes can obtain that image more rapidly, because of its superior light-gathering power, and in addition can secure images of objects too faint to be revealed by a smaller telescope.

To achieve improved angular resolution, as well as to observe the cosmos in ultraviolet radiation, the scientific community, with a crucial assist from taxpayers in the United States and abroad, sent the Hubble Space Telescope above the atmosphere. The HST has a mirror just 2.4 meters wide—the largest that could fit within the budget and the space shuttle that launched the HST. Puny though this may be in comparison with the mighty Kecks, the HST's location compensates mightily. At a wavelength of 60 millionths of a centimeter, the HST's theoretical best angular resolution equals 250,000 times ($6 \times 10^{-5}/240$), or 0.0625 second of arc. Amazingly enough, thanks to its precise guiding system and its refurbished optics, the HST can now achieve nearly this angular resolution, which provides by far the sharpest pictures of cosmic objects (Color Plate 7).

Can the Hubble Space Telescope Observe Planets around Other Stars?

At how great a distance could the HST resolve Alpha Centauri A and B as separate objects? Since Alpha Centauri A and B are separated by about 18 seconds of arc as we see them at their distance of 4.4 light years, and since this angular separation would decrease in direct proportion to the distance, an increase in the distance by a factor of 300 would reduce the angular separation to 0.06 second of arc. Thus the HST could see Alpha Centauri A and B as individual objects out to a distance of nearly 1,300 light years. This still amounts to only about 1 percent of the distance across the Milky Way galaxy, though millions of stars lie within this distance from the sun.

Now imagine replacing one of the Alpha Centauri stars with a planet. After all, the 23-A.U. separation of the two stars corresponds roughly to Jupiter's distance from the sun (5.2 A.U.), to Saturn's (9.5 A.U.), or, even better, to Uranus's and Neptune's distances (19.8 and 30 A.U., respectively). Couldn't we see this conceptual planet with the Hubble Space Telescope, even if the planet-star system had distances as large as a thousand light years?

To this we must answer with a resounding *No! Certainly not! Impossible!* Why this clamor? We must emphasize that the criterion stated above for distinguishing two objects as individual sources applies *only if the objects have equal brightnesses*. The greater the deviation from equality, the more difficulty we face in spotting the fainter object next to the brighter one. You can see this for yourself, if you and a friend perform an experiment with flashlights of various brightnesses held close together, or if you imagine trying to see a firefly close to a searchlight. Whether or not you engage in these activities, the rule remains, and all astronomers know it well. Even a brightness contrast by a factor of 5 or 10 will make spotting the faint object far more difficult than distinguishing two objects of equal brightness. In the case of extrasolar planets, a factor of 5 or 10 barely begins to suggest the actual situation.

In reality, because planets shine only with the light they reflect from their stars, the typical contrast in apparent brightness between a star and one of its planets amounts to about a billion to one! If, for example, you visited the Alpha Centauri and looked back at the solar system, the sun would have an apparent brightness 600 million times Jupiter's, 2.4 billion times Saturn's, 20 billion times Uranus's, and 25 billion times Earth's. Note that a trade-off occurs: Planets closer to the sun receive and reflect more sunlight per square meter, so that, for example, seen from a great distance, Uranus would appear brighter than Neptune, even though the two planets have similar sizes. On the other hand, planets farther from a star are more easily detectable, because the enormous interference from the star's own light decreases with distance. All in all, though, the situation seems far from promising; the factor of nearly a billion for Jupiter means that a Hubble Space Telescope in orbit around a planet in the Alpha Centauri system could detect neither Jupiter nor any other of the sun's planets.

During the early 1970s, while the HST was under design and construction, its creators added a *coronagraphic finger* to help the telescope search for extrasolar planets. This term, another fine example of how astronomers (and others too) transpose the familiar into less-familiar situations, refers to a *coronagraph*, a device invented early in this century to allow astronomers to observe the sun's gauzy outer halo, called the corona. Because the corona's glow only equals the full moon's, the light from the sun's disk, nearly a million times more intense, completely overwhelms the pale fire of the corona. Only during a total solar eclipse, when the moon completely covers the solar disk, can we see the solar corona with our unaided eyes. The coronagraph places an opaque disk in the line of sight, with exactly the same angular size as the sun, and thus allows the innermost, brightest part of the corona to become visible, provided that the atmosphere remains still. (Never try, as some have, to make an amateur coronagraph by holding a coin at arm's length in front of the sun; the system will not work, because your arm is too flexible, and you risk blinding yourself.)

The HST's coronagraphic finger aimed to put a small disk in the line of sight to a star, blocking the starlight but covering only a small part of the field of view, so that planets orbiting the star could be detected. Provided that the telescope's guidance system could maintain the proper alignment as the HST orbited the Earth at thousands of kilometers per hour, this offered the possibility of making direct observations of Jupiter-like planets around the closest stars. As things turned out, however, the coronagraphic finger was never deployed. The main mirror's incorrect curvature meant that far more time than planned had to be spent observing individual objects, for the astronomers had to collect extra amounts of light to process with computers, in order to achieve an angular resolution close to the theoretical best. In this situation, use of the coronagraphic finger would have consumed unallowably large amounts of observing time. When the astronauts installed the HST's corrective optics in December 1993, they removed the part of the optical system that contained the coronagraphic finger, which therefore never saw action in space.

NASA plans an upgrade mission to the HST in 1999 that will install a new camera with a coronagraphic finger that will allow the telescope to look for planets around nearby stars. The billion-to-one brightness ratio still implies that these planets would lie just at the edge of detectability with the HST. However, as astronomers improve and extend their Doppler-shift and other techniques for finding planets, they will soon have several cases in which Jupiter-like planets orbit stars at Jupiter-like distances. (They already have three cases that approximate that situation, the planets around 47 Ursae Majoris and 16 Cygni B and the second planet around 55 Cancri, in which Jupiter-like planets orbit at distances comparable to Jupiter's distance from the sun.) These would provide the first targets for the upgraded HST. Eventually, astronomers hope for even better optical performance and hence an even greater ability to see extrasolar planets directly, from the Next Generation Space Telescope (NGST) described in chapter 9, which would have a mirror several times larger than the HST's. For now, the NGST remains a gleam in the eyes of forward-looking astronomers, but if the public (through its elected representatives) so chooses, this telescope may yet fly.

In that case, the younger members of society may someday enjoy scanning images of stars with planets around them. We must emphasize, however, that these images will show any planets as a single point of light, if they reveal them at all. This hard fact clashes with our desire to study planets in detail, to admire their continents and oceans (if they have them), their polar caps and deserts (should they exist), their markings, rotations, nighttime lights of civilization (if those are present). The desire for *pictures* has driven much of the interest in the search for planets, both from the public and within NASA, which, in times of increased budget difficulties, has begun to search more diligently for what the public seems to want. In May 1994 Dan Goldin, the head of NASA, stated that "in the not too distant future, we will have the technology needed to image any planets that might orbit nearby stars. . . . Perhaps, just perhaps, the next generation's legacy will be an image of a planet 30 light years from Earth—and the tools of technology it took to capture that image, all combined in one gift to our children, 25 years from now." By using the word *image* rather than *picture*, Goldin blurred a distinction most of us make automatically. To an astronomer, an *image* can have any range of detail, from a single point to the kind of clarity shown in Color Plate 7. To the public, however, and probably to Goldin as well, an *image* in this context means a view that shows a recognizable planet. By January 1996, having learned somewhat more about what it would take to produce this sort of image, Goldin stated that he "wanted to get the ball in play," meaning to arouse public interest in the search for extrasolar planets. "We shouldn't promise the public a picture," he added. "That's what gets us into trouble—promises you can't deliver on."

For Pictures of Planets, Build an Interferometer

The rules of optics proclaim that an observational instrument can achieve higher angular resolution by either decreasing the wavelength or increasing its light-collecting diameter. However, this rule does not require that the light collector fill the entire region within its span. If, for example, we remove the central portions of a telescope mirror, leaving only a circular band around the mirror's circumference, the telescope cannot gather as much light as it did when whole, but it will have nearly the same angular resolution for the light that it *does* collect. The same holds true if we take away most of the circular band, leaving only two collecting areas on opposite sides of the original mirror. We may thereby decrease the original mirror's light-collecting ability by a factor of a thousand, but the optical system will maintain almost the same angular resolution. This means that we shall lose the ability to detect faint sources at all, but we shall be able to see the brighter ones with the same fineness of angular detail as before, although the sources will appear much fainter.

Astronomers call a system that combines observations from separated wave collectors to produce a single image an *interferometer*. The name refers to the fact that in order to combine the observations, an interferometer must compare the waves collected by its components, allowing the waves to "interfere" with one another (Color Plate 8). In the dissected mirror described above, the interference process occurs when the remaining areas of the mirror focus light waves to a single point. With great ingenuity, astronomers have created other, more complex means of arranging for interference to occur. In the most prominent and successful examples, radio astronomers have created interferometers consisting of arrays of individual radio dishes. The astronomers record the observations made by each dish, along with accurate timing signals. A central computer then processes the results from each dish, calculating how the different simultaneous observations interfere with one another. From this processing, the computer calculates a single image, whose angular resolution equals that attainable by a single dish with a diameter equal to the maximum separation of the dishes.

One of the most productive radio interferometer systems, the Very Large Array, or VLA, spreads over more than twenty miles on the high plains of central New Mexico. There, twenty-seven individual radio dishes move along railroad tracks to create interferometers of different sizes, capable of producing detailed maps of the radio emission from cosmic objects (Color Plate 9). Radio astronomers have extended the technique of interferometry to much larger distances by recording signals simultaneously at radio observatories around the world and then comparing these signals to see how the wave patterns interfere with each other. In this way, they have, on occasion, created a radio telescope effectively as large as the Earth!

Radio astronomers require interferometric techniques to observe sources in detail, because radio waves have wavelengths that typically exceed those of visible light by a factor of about one hundred thousand. The rule we have met then implies that in order to match the angular resolution of a visible-light telescope, a radio telescope must have a diameter one hundred thousand times greater. Impressively, the "radio telescope as large as the Earth," with a diameter about a million times larger than the Keck Telescope's mirrors, meets and surpasses this challenge. As a result, radio interferometers have secured images with even finer angular resolution than *any* picture taken by the Kecks or the Hubble Space Telescope.

Creating an interferometer for radio waves has proven to be far easier than making one for visible-light waves, for technical reasons that hinge on the much shorter wavelengths of light waves. At present, several visible-light and infrared interferometer systems are under development. We shall discuss them in more detail in chapter 9 and may note now the general conclusion that within ten or fifteen years Earth-based interferometer systems may be able to discern planets the sizes of Jupiter and Saturn, with 11 and 9 times the diameter of Earth, respectively, if such planets exist around stars within a few dozen light years of the sun. These interferometers may even prove capable of seeing planets like Uranus and Neptune, with diameters about 4 times Earth's, orbiting at 20 to 50 A.U. from their stars. But in order to see Earth-like planets, orbiting at a distance comparable to 1 A.U. from a sunlike star a few dozen light years from the solar system, we must plan on sending interferometers into space, once we have developed and proven the interferometer technology we shall require.

Meanwhile, we may note with some regret that the natural evolution of optical systems on Earth has not led to interferometers. Too bad! If our brains could arrange for the interference of light waves that our two eyes detect, we could achieve the same angular resolution as that of a single collecting system with an effective diameter equal to the distance between our eyes. This system's angular resolution would be about 30 times better than a single eye can provide, since the interferometer would span our interocular span, about 7.5 centimeters, instead of the 0.25 centimeter of each eyeball lens. Then we would view the world with an angular resolution of 2 seconds of arc! Each of us would, in effect, have a telescope with a 7.5-centimeter-wide lens at our disposal. This would allow us, when observing a suspension bridge from a distance of ten miles, to see not only the three-foot-wide main cables but also small supporting wires, not much more than an inch in diameter. We could easily recognize one face from another at a distance of a mile! Other animals, with greater interocular distances, might see the world with the clarity provided by a professional telescope.

Evolution turned us out otherwise, and no animal has interferometric capacity in its binocular vision. Although our brains do use bifocal vision to make distance estimates, they do not, and cannot, use the interference of light waves detected by each eye to create a picture of the world around us. But what nature has not done, humans can create and can improve. Radio interferometers work on Earth, where the atmosphere only barely affects the passage of many wavelengths of radio waves. For visible-light interferometry, space beckons, since we have already encountered the limiting effects of Earth's airy veil. Chapter 9 will pick up the dreams of the future, in which space-borne interferometers provide pictures of planets around other stars. This will reward us for examining other key aspects about the search for planets, starting with the nagging question, How do planets form in the first place?

The Pleiades, 400 light years form the solar system, are a cluster of several hundred stars that began to shine about 50 million years ago, so recently that wisps of the gas from which the stars formed can still be seen around them.

CHAPTER

4

The Formation of Planetary Systems

Planet Formation: Theory and Evidence

In an ideal world, we would know not only what we are but also how, when, and why we came to be this way. However, our own history demonstrates that our exploration of the influences that made us what we are poses even greater difficulties than comprehending the world around us. Though traces of the past lie all around us, reading the evidence unambiguously never proves easy. Likewise for planets: We live on Earth, observe the solar system billions of years after its formation, and have only this single example for detailed examination. Nevertheless, to paraphrase Mark Twain, astronomers have built impressive edifices of speculation from a few pieces of hard data about the early solar system. As they now gather rudimentary information about other planetary systems, they must revise and expand their theories, thus advancing scientific progress while losing the false sense of certainty that cosmic generalization from a single example provides. The testing of theory against fact, an action that distinguishes science from most other human activities, should soon lead scientists to a better understanding of how planets formed around the sun and its neighbor stars in the Milky Way galaxy.

The Relics of Planet Formation

Recovering the past requires relics, objects so little altered by the passage of time that they still furnish evidence about bygone eras. Because time wears all things away, finding these relics requires serious effort. On our planet, erosion and plate-tectonic motions have worn away and buried all the rocks that first formed our planet's surface, replacing them with other rocks that no longer carry a record of what happened near the time of origin. Dating rocks by the decay of radioactive elements they contain has shown that the oldest rocks found on Earth's surface, from the Isua shelf in Greenland, have ages close to 3.8 billion years. Surpassing this record, the ages of many of the rocks collected in the highlands of the moon fall between 4.2 and 4.4 billion years. But the oldest "rocks" of all are the primitive meteorites, also called *chondrites*, a class of meteorites that usually contain small, round inclusions, about a millimeter across, called *chondrules*. The primitive meteorites have ages between 4.43 and 4.55 billion years. Planetary astronomers believe that the chondrules condensed from the primordial material that made the solar system; they would like to verify this conclusion by getting their hands on a comet. Comets are thought to be the oldest and (in most cases) the least altered relics of the solar system, frozen since time began in the solar system. So far, the closest astronomers have come to their elusive goal amounts to the observations made during rapid passes by Halley's Comet in 1986, when the comet last came relatively close to Earth (Color Plate 10).

From these data, specialists in planetary formation conclude that the entire solar system—the sun and its planets, satellites, comets, asteroids, and meteorites—formed about 4.55 billion years ago. Wait a minute! From the ages of a few rocks fallen from interplanetary space, astronomers presume to derive the age of the *sun*? This would be a non sequitur indeed, if astronomers did not have a theory that places the formation of the entire solar system within a single epoch of time. They do have such a theory, which also provides the mental framework from which to address the newly discovered planets.

The Standard Model: Star and Planet Formation Together

During the past five decades, astronomers have created what we may call the *standard model* of star formation, which begins when a region within an interstellar cloud of gas and dust increases in density to the point that it begins to contract. Just what causes this increase in density remains an astronomical mystery, shrouded within the opaque veils produced by interstellar dust grains (Color Plate 11). Perhaps a nearby supernova explosion ruffles the cloud slightly; perhaps gravitational forces among interstellar clouds perturb their inner structure. Once a region within the cloud grows somewhat denser than average, it will continue to contract through *self-gravitation*, as each part of the contracting region gravitationally attracts every other part. The start of the process has proven far more difficult to understand than the later phases, which can be well modeled on computers. However, this works only by comparison; to state that astronomers completely understand these later phases would be a serious exaggeration.

Thus the early stages of star formation take place in a roughly spherical region, a small cloud containing interstellar gas and dust, that shrinks to a smaller size by self-gravitation. If the cloud happens to have some rotation, as most of these small clouds apparently do, then the later stages of contraction will be favored to occur in directions parallel to the rotation axis rather than perpendicular to it (Color Plate 12). This favoritism arises from what scientists call the *conservation of angular momentum*, the fact that the rotation rate increases as the object shrinks in size in directions perpendicular to the rotation axis. High divers use this fact to spin more rapidly when they contract their bodies, less rapidly when they extend to full length. A contracting interstellar cloud does shrink in all three directions, but because this contraction occurs more readily along the spin axis than in the two perpendicular directions, it assumes a pancakelike shape, greatly flattened along its spin axis.

Just what happens within that rotating pancake determines whether the cloud will produce a single star or a double- or multiple-star system, and whether or not much smaller objects such as planets will accompany the formation of the star or stars. When astronomers survey the Milky Way, they find that about half of all the stars they can examine closely turn out to be double, triple, or even higher multiples, with double stars the most common of these. Hence most of the stars in our galaxy are not singletons but instead were born with companion stars and move in orbit with them around the center of mass of the double- or multiple-star system. A recent success in modeling star formation with computers shows that double stars are favored to form in most condensations.

Whatever the stellar outcome, the standard model predicts that stars form from a rotating pancake of gas and dust, which develops particularly dense regions that contract to form *protostars* (stars-in-formation) and then stars. The remainder of the pancake forms a *protoplanetary disk*, a flattened aggregation of matter that holds the key to making planets. As the protoplanetary disk initially forms, it includes both gas and dust. The gas consists primarily of hydrogen molecules (H_2), along with hydrogen and helium atoms, plus other simple molecules such as carbon monoxide (CO), carbon dioxide (CO_2), nitrogen (N_2), methane (CH_4), ammonia (NH_3), and water vapor (H_2O). The dust grains, each about 1/1,000 of an inch across, are clumps of a million or so atoms, most of them carbon, silicon, and oxygen; they resemble tiny footballs made of graphite and silicates, with outer coatings of water ice or frozen carbon dioxide (dry ice).

Plenty of Protoplanetary Disks in the Milky Way

Throughout the Milky Way, protoplanetary disks exist in abundance. More precisely, astronomers possess a large body of observational evidence that both young stars and objects considered likely to form stars are surrounded by rotating, flattened disks of material. This does not prove that the disks *do* form planets, even though we may name them "protoplanetary disks." To planet-formation experts, however, the observational situation appears to confirm the standard model—at least up to the point where protoplanetary disks have formed.

Astronomers know from both direct and indirect observation that protoplanetary disks exist. In some instances, they have actually photographed the material in a disk (Color Plate 13). In many more cases, astronomers have detected radio, submillimeter, or infrared waves emitted by material surrounding a star. This material has a roughly flattened shape, and Doppler-shift measurements of its emission demonstrate that each disk *rotates*, with the outer parts moving more slowly than the inner parts, in just the manner expected when material moves in orbit around a massive object at its center. A typical disk extends to distances of about 1,000 A.U. from its star, but the material within 100 A.U. has a noticeably higher density than the remainder. In some highly suggestive cases, such as the disk around the star Beta Pictoris (Color Plate 13), the innermost parts of the disk constitute a clear zone that appears to be almost devoid of gas and dust. Since planets are far more difficult to observe than material in a protoplanetary disk, the *absence* of material in the disk suggests the *presence* of planets: Astronomers find a situation that matches what they expect if the matter if the disk's inner regions has turned into planets.

Beta Pictoris represents an unusual case: a mature star that still possesses a disk of matter rotating around it. Astronomers have used the Hubble Space Telescope to make detailed studies of this disk, which extends about fifty times the Earth-sun distance outward from the star. These studies have revealed a warp in the disk's inner regions, so that the disk's innermost parts orbit the star in a plane that is tilted with respect to the rest of the disk. Because the gravitational force from the rest of the disk tends to pull the warped part into the same plane, the astronomers conclude that something keeps the warp in existence. That something might be the gravitational effect from a passing star, but it seems more likely to be the continuing pull from a planet orbiting Beta Pictoris inside the disk. Christopher Burrows of the Space Telescope Science Institute (located on the Johns Hopkins University campus in Baltimore, Maryland, and in charge of the scientific component of the HST's operation), estimates that to maintain the warp, this planet must have a mass from one-twentieth to twenty times Jupiter's mass. The planet might be orbiting within the clear zone or might move through the inner, tilted parts of the disk. In 1997 the Hubble Space Telescope will receive a camera system that can record infrared radiation. Since the disk radiates primarily infrared, this will allow the HST to make more detailed observations of the Beta Pictoris disk than are now possible; these observations should provide us with key information about the details of the disk that can tell us more about what causes the warp in its inner regions.

In contrast to the Beta Pictoris disk, which orbits a mature star, most protoplanetary disks appear around young stars, and in particular around a class called T Tauri stars, identified as extremely young by the fact that they lie close to star-forming clouds. The spectra of the light from T Tauri stars can be best explained as the combination of visible-light emission from a newly formed star plus the infrared emission from a disk of material heated to roughly "room temperature" by the star's radiation. Most of the T Tauri stars have protoplanetary disks, a fact established both by ground-based observations of many dozen objects and by an analysis of a much more numerous host of infrared-emitting objects detected in the survey made by *IRAS* (*InfraRed Astronomy Satellite*). Although *IRAS* could not examine any of these objects in detail, the experts judge that the excess infrared emission from many of these sources most likely arises form dusty material in orbit around young stars.

The combined observations of individual disks, T Tauri stars, and the *IRAS* infrared survey suggest that our galaxy contains millions of protoplanetary disks. How can it be that disks are so common, and planets so rare? Astronomers now feel sure that the paradox arises from false logic in the question. Planetary systems should prove to be at least as abundant, probably more so, than protoplanetary disks. *The difference lies in the relative ease of detection.* Because matter in a disk spreads through a wide volume, each part of the disk can emit infrared radiation that escapes into space. In contrast, planets "hide" most of their matter within themselves, where it cannot produce radiation that an observer might detect. If we disassembled the Earth, we could put a disk of matter in orbit around the sun whose infrared glow would be hundreds of thousands of times greater than the actual Earth's. We must struggle so hard to find planets precisely because the disks that made them packed once-rarefied material into dense spheres, hiding it from easy discovery. Calculations imply that the disk phase of planetary formation lasts for only a small fraction of a planetary system's total lifetime, though some stars may maintain disks of matter in orbit around them for a billion years or more, perhaps never forming any planets. If the protoplanetary disks around most stars, or at least around most single stars, do turn their matter into planets within, say, a few hundred million years, then (since the stars themselves last for many billion years), we expect planetary systems to outnumber protoplanetary disks by a factor of ten or more. Because astronomers know the relative ease of finding planets and protoplanetary disks, the discovery of additional planetary systems should soon allow them to establish this ratio with greater accuracy.

How Do Protoplanetary Disks Make Planets?

A rotating protoplanetary disk has the "job" of turning its outer regions into planets while its inner parts become a star or stars. Many disks may never perform this feat, but, as the recent discoveries demonstrate, many have succeeded. The standard model envisages three key stages in the planet-formation process. First, material in the rotating disk agglomerates to form objects a few inches, or perhaps a few feet, in diameter. Second, these rock-sized objects collide to make *planetesimals*, objects up to a kilometer or so in diameter. Third, the planetesimals interact to make planet-sized objects. We divide the total process into three phases because important differences exist in the ways that particles interact during each phase.

The first phase sees the dust grains within the protoplanetary disk "raining" toward the plane of the disk, pulled by the collective gravitational force from the disk. This concentrates the dust grains, allowing them to collide and to stick together more rapidly in a process of coagulation that lasts for several thousand years. The coagulation produces a host of pebble- and rock-sized particles, all orbiting the protostar at the center in the same direction and in nearly the same plane. A close analog to this situation exists today in the rings of Saturn, where particles of similar sizes likewise orbit a massive object, all in the same direction, within a region only a few hundred yards thick.

In the second phase, gravitational interactions among the individual pebbles and rocks produce instabilities that concentrate the objects into particular orbits, with nearly empty gaps between the crowded orbits. If the swarm originally had a completely even distribution of its objects among all the possible orbits, the individual objects could quite likely maintain their orbits for billions of years. If, however, the total mass of the objects in some of the orbits exceeds the average, those orbits will grow progressively more populated, drawing objects from nearby orbits by the modest gravitational forces produced by orbital overpopulation. To envision the solar system at the end of stage two, imagine a trillion of these "overpopulated" orbits, with an equal number of gaps between them; most of the orbits are circular, but some have elongated, elliptical shapes. Within each overpopulated orbit, the matter eventually produces a single planetesimal roughly the size of a small mountain. In Saturn's rings, phase two cannot occur because of the combined gravitational effects of Saturn and its innermost satellites (which are themselves the product of all three stages, operating within the solar system at earlier epochs.)

Phase three involves the collision of planetesimals to make planets. The mutual gravitational interactions of the planetesimals make them change their orbits slightly as time passes; many of these orbital changes produce collisions. Although some collisions would have fragmented a sizable planetesimal into much smaller pieces, most of them occurred at speeds that produce a single, larger object from two planetesimals. Within a few thousand years, according to calculations supporting the standard model, most of the individual planetesimals would combine to yield a few thousand objects, each a few hundred miles in diameter.

Here we leave phase three for the endgame of planet formation, "phase three and a half," in which the massive planetesimals attract one another by gravitational forces, eventually collide with one another, and produce—what? In the case of our solar system, we think that the collisions gave birth to nine planets, seven large satellites (each many times larger than the largest planetesimals), plus a few dozen satellites, a host of asteroids and comets, and a modest amount of interplanetary gas and dust (Color Plate 14). The sizes of asteroids and comets, which range from a few hundred miles to less than one mile, suggest that these are planetesimals that grew to modest size without joining larger objects. The smallest objects in the solar system, the swarms of rock-, pebble-, and sand-sized meteoroids, are presumably debris left over from the end of stage one of the standard model. In contrast, the sixteen largest objects in the solar system must have formed from numerous collisions among planetesimals.

The giant planets each have so much mass, mostly in the form of hydrogen and helium gas, that we must add two assumptions to explain their formation: that they managed to form sizable solid cores (estimated at 10 to 30 Earth masses for Jupiter, with slightly less massive cores for Saturn, Uranus, and Neptune), and that these cores then attracted much larger amounts of hydrogen and helium, which had never participated in the coagulation that formed planetesimals. The planets forming in the inner regions of the solar system failed to attract and retain much gas, for two reasons. First, they received far more heat from the protosun than planets 10 to 100 times farther away, which prevented ice from forming and made the gas more difficult to retain. Second, the inner regions of the solar system contained far less material to begin with, because the outer regions of the disk included a much greater volume than the inner ones did. Hence the protoplanetary objects that became Mercury, Venus, Earth, and Mars could never "grow" massive rock-and-ice cores that could attract and retain the large amounts of hydrogen and helium providing the bulk of Jupiter, Saturn, Uranus, and Neptune. Even in the outer solar system, this attraction and retention requires cores many times more massive than Earth's, and this material was simply not available to the inner planets as they formed. The escape of gas from the inner parts of the disk, once the protosun began to shine, thus seems entirely natural. The giant planets' capture and retention of hydrogen and helium appears to be the equally natural result of having sufficiently large amounts of material, and sufficiently low temperatures, to form large cores of rock and ice.

Other scenarios for stage three and a half seem entirely possible. We can imagine that one object formed from planetesimal collisions has far more mass than any other and embarks on a "runaway" that embraces essentially all the matter in the disk, leaving the star with a single planet. At the other end of the scale, a number of planetesimals might grow at roughly equal rates, so that a star could have, say, two dozen rocky planets, each with a mass between 0.1 and 10 times Earth's mass. This could occur if a protostar's heat evaporated the gas in the disk before any massive core could attract it to produce a giant planet. Indeed, avoiding this no-giant-planet result, as our own planetary system managed to do, poses a difficulty to planet-formation theorists.

Stage three of the formation process lasts for a few tens of millions of years, during which the star emits steadily greater amounts of energy (see below), and the stellar wind—elementary particles expelled from the star at high velocities—blows gas away from the star into interstellar space. The matter that does not evaporate or blow away slowly spirals inward toward the star, and the completion of the formation process finds the planetary region essentially empty—except for the larger objects that have formed. Because interplanetary space is so large, however, a huge number of objects have ended up in orbit around our sun, and presumably around other stars too (see Color Plate 14).

If we leave the big picture to focus our gaze closer to home, we may note that the most widely accepted theory for the formation of our moon assigns this event to a giant impact, during stage three and a half, when a Mars-sized object struck the proto-Earth and knocked loose a mass of material that coagulated to form the moon. This explains why the chemical composition of the moon so closely resembles the Earth's, without being absolutely identical, and how the Earth came to have such a massive satellite: The moon has 1/81 of the Earth's mass, while no other satellite has even 1/1,000 of its planet's mass—except for Charon, the single moon of Pluto, which has fully half Pluto's diameter and more than 10 percent of its mass. Since Pluto itself is smaller and less massive than our moon, and since the compositions of both Pluto and Charon resemble the nuclei of comets, the Pluto-Charon system probably consists of two primordial, icy objects that happened to acquire orbits around their common center of mass. In the case of our moon, astronomers now place greater reliance on an explanation based on impact rather than on capture.

Phases one, two, and three (including the different scenarios in stage three and a half) should have taken a total of a few hundred million years. Stages one and two proceed in no more than a few million years, but the collisions of planetesimals take rather longer, since the orbital gaps between them have grown to considerable size. The final stages of planet formation—at least of Earth-like planets that lack the gaseous layers that envelop the giant planets—consist of a rain of bombardment, in which the last chunks of matter that form the planet could be seen (by whatever was there to see them) as dangerous incoming missiles. Even today, at intervals of tens of millions of years, a small planetesimal (asteroid) with an orbit that happens to cross the Earth's will actually strike our planet, raising the havoc that can produce "mass extinctions" such as the one that ended the reign of the dinosaurs, 65 million years ago.

The Spacing of Planetary Orbits

Astronomers developed the standard model for the formation of planetary systems with one basic model in mind: our solar system. Not surprisingly, then, the standard model does a good job of explaining the basic characteristics of the planets that orbit the sun. The flat protoplanetary disk explains why the planets all orbit the sun in the same direction and in nearly the same plane. Computer models of stages two and three, the formation and interaction of planetesimals, agree fairly well with another basic orbital feature of the sun's planets: The gap between successive orbits approximately doubles each time we look one more planet outward, provided that the asteroids, which orbit at about 2.8 A.U. from the sun, are collectively considered to be a planet. A good case can be made for the asteroids as the remnants of a "failed planet," one that could never form from planetesimals because Jupiter's gravitational forces perturbed the asteroids' orbits in a way that scattered most of the material in this region and kept the asteroids from colliding. Even though the spacing law begins only with the gap between Venus and Earth and breaks down beyond the orbit of Uranus, it works well for six successive gaps: Venus-Earth, Earth-Mars, Mars-asteroids, asteroids-Jupiter, Jupiter-Saturn, and Saturn-Uranus, with spacings (in A.U.) of 0.28, 0.52, 1.28, 2.4, 4.4, and 9.6, respectively. Neptune lies at a distance of 10.8 A.U., and Pluto at 20.2 A.U., beyond Uranus.

Building the Planets: Capture versus Escape of Material

Four billion five hundred million years ago, according to the meteoritic evidence and astronomers' best models for stellar evolution, our sun began nuclear fusion in its core and passed from protostar to star status. If the protosun had not produced heat, this would have marked the time when the protoplanets forming around it first reacted to solar heating. In fact, however, the protosun had been growing ever hotter for the previous 50 or 100 million years, so its heat had already affected its protoplanets. The standard model of planetary formation envisions a contest throughout the protoplanetary disk between protoplanets' ability to attract gaseous material gravitationally and the protostar's (and then the star's) ability to evaporate that gas and to blow it away with a wind of fast-moving particles. Evaporation occurs as heat from the protostar warms the gas, while the stellar wind, made mainly of protons and electrons, expels gas molecules one by one as collisions occur. Even today, the solar wind blows about one part in a hundred trillion of the sun's total mass into interstellar space each year. Four and a half billion years ago, when the sun first began to shine by nuclear fusion, this rate of mass loss was probably higher, as it seems to be for many of the young stars that astronomers observe.

The competition between the accumulation and loss of material has left the sun's planets as we find them: four small, rocky, inner planets; four large, mostly liquid, outer planets; and Pluto, smaller than our own moon and more akin to a giant comet than to any of the other planets. These differences find ready explanation in the standard model: The four inner planets lost the competition to hold onto the gas that once enveloped them, because they failed to grow sufficiently massive before the gas escaped from the inner solar system, while the next four planets outward won the battle. Around their solid cores, the giant planets' enormous self-gravitational forces squeeze hydrogen and helium, which provides the bulk of these planets' masses, to the point that they become mostly liquid, with gaseous layers above the liquid that form the planets' "surfaces." The ninth planet, Pluto, can be designated an oddball, a cometlike object, resembling the icy satellites of Jupiter, Saturn, Uranus, or Neptune more closely than any of the other planets.

Notice that *retaining* gas proves easier than *acquiring* it, because the mass of the gas itself increases the planet's self-gravitation. If we could move Jupiter into Earth's orbit, we would find that Jupiter lost almost no mass, despite a rise by a couple of hundred degrees in the temperature of its outer layers. However, Jupiter could never have *formed* at the Earth's distance from the sun, where the loss of gas mass prevailed over gravity's attempts to hold it. Our own atmosphere apparently arose from two processes: outgassing from volcanic vents that released gases trapped in rock layers; and bombardment by cometlike planetesimals, rich in frozen ammonia, water vapor, and carbon dioxide, that smashed into the proto-Earth as it ended its formation

Today, billions of years after the sun began to shine, interplanetary space contains none of the gas that once formed the bulk of the protoplanetary disk. Journeys between the planets carry a small risk of high-speed collisions with some of the dust- and pebble-sized debris left over from the epoch of formation, but the sun's heat and the solar wind have essentially cleared out the once-mighty disk. (Everything is relative, of course: Most of the mass in the disk joined the protosun, where we can still enjoy it.) The standard model predicts that events around other protostars should have proceeded in a similar fashion, at least so long as the bulk of the disk's mass formed a single star. We can well imagine that other stars formed with dense, rocky inner planets and giant outer planets and even with giant planets possessing satellite systems that resemble Jupiter's and Saturn's. There, in miniature variants of the solar system, we find that a disk of rotating matter apparently formed a series of moons, which all orbit the planet in the same direction and in nearly the same plane. In the case of Jupiter's four large and massive satellites, we even find a rough division into inner and outer sets: The two inner moons, Io and Europa, have significantly higher densities than the outer two, Ganymede and Callisto. Planetary astronomers conclude that Io and Europa consist mainly of rock, while Ganymede and Callisto are each about half ice and half rocky material, much like the composition of the cores of the giant planets. Since Jupiter creates some heat from its own slow contraction, we may speculate that Ganymede and Callisto did better at retaining some of the lighter molecules around them than did Io and Europa, which received more heat from their planet.

One Prediction of the Standard Model: Nearly Circular Orbits for Large Objects

The standard model for the formation of a planetary system envisages a protoplanetary disk of gas and dust, within which clumps of matter form from dust particles moving in nearly circular orbits. The model predicts that planetesimals should likewise have nearly circular orbits, because each one forms within a small zone characterized by a particular distance from the center. Even though the planetesimals must deviate somewhat from these orbits as they attract one another and collide, the planets they produce, the result of many such collisions, should have orbits for which most of the deviations have canceled one another and therefore have nearly circular shapes. Each orbit may well be slightly elliptical, but the orbit's *eccentricity*, a measure of its deviation from circularity, will be small. In the solar system, all the planets except Mercury and Pluto have only slightly elongated orbits. The orbital eccentricities of the other seven planets, which equal the differences between the planets' maximum and minimum distances from the sun, divided by the length of the long axis, all fall below 0.1. Even for Mercury and Pluto, the orbital eccentricities equal 0.21 and 0.25, respectively, which makes the planets' maximum distances from the sun 53 and 67 percent greater than their minimum distances. When we compare these orbits with the orbit of Halley's Comet, which has an orbital eccentricity of 0.97 and a maximum-to-minimum distance ratio of 66, we can see that the sun's planets have only modest eccentricities, compared to the full range of possibility.

Of the eight new planets described in chapter 1, all but two have orbits with eccentricities of 0.03 or less (see Table 1.1). If the two exceptions—the planets around 70 Virginis and 16 Cygni B—did not exist, we could state that the standard model has received confirmation from the new discoveries, albeit based on few numbers. If we truly have confidence in the standard model, we could say that the large eccentricities (0.40 and 0.57) for the orbits of the planets around 70 Virginis and 16 Cygni B argue that these objects are not planets. As we discuss below, these objects might be brown dwarfs rather than true planets. However, experience has shown that models based one or two examples carry only modest weight. We ought to study a number of planetary systems—a dozen, at least—before claiming a high degree of confidence in our generalizations about the ways that planets form and the orbits that they occupy.

The Kuiper Belt and the Oort Cloud

In addition to the large objects that orbit the sun in nearly the same plane and in the same direction (the planets and their satellites, as well as the larger asteroids), the solar system contains a host of comets and meteoroids. The *comets* are "dirty snowballs," typically a few kilometers in diameter, made of ice, dry ice, frozen alcohol, and dust grains; the meteoroids are rock or rock-and-metal objects, ranging from a few hundred meters across down to submicroscopic sizes. *Meteoroids* have essentially random orbits, with all possible directions and orientations with respect to the plane of the planets' orbits. Comets fall into one of two categories, based on their orbits, each of which bears the name of a different famous twentieth-century Dutch astronomer. The comets in the Kuiper belt, named after Gerard Kuiper, orbit the sun in roughly the same plane that the planets do, but at distances ranging from about 50 A.U., somewhat greater than Pluto's, out to hundreds or even a few thousand A.U. Collectively, the Kuiper-belt comets form a vast, relatively flat system of cosmic snowballs that may well be the remnant of the sun's protoplanetary disk. In contrast, the Oort cloud of comets, named after Jan Oort, contains comets of all orbital orientations and extends out to tens of thousands of A.U.

Current theories of the Oort cloud assign its origin to the wholesale gravitational scattering of comets, early in the history of the solar system, by the giant planets. When a cometary planetesimal happened to pass close to one of the giant planets, the planet's gravity might have captured it, adding it to its already considerable mass, or it might have exerted a "gravitational slingshot effect" that flung the comet into a huge orbit, hundreds of times larger than any planet's orbit around the sun. Uranus and Neptune played the key role in this process, because Jupiter and Saturn have such large masses that they would usually send a comet not into the Oort cloud but rather far beyond it, into the depths of interstellar space! Even though the solar system must have lost immense numbers of comets in this way, the more modest scattering by Uranus and Neptune left the Oort cloud with an estimated trillion or so comets. The comets' combined masses, however, probably do not equal many times the Earth's mass and fall far short of any giant planet's mass. In contrast, the Kuiper belt comets are the remnants of the original protoplanetary disk that formed the solar system.

Do the Giant Planets Protect Life on Earth?

We might find this story only an interesting digression, were it not for the hypothesis, proposed first by George Wetherill of the Carnegie Institution of Washington, that the creation of the Oort cloud allowed life to exist on Earth. The bombardment of the early Earth by planetesimals seems well established, both by the theory embodied in the standard model of planetary formation and by the geological record embedded in our nearby moon. In addition, periodic bombardments by planetesimal objects—comets or asteroids—appear at intervals of 10 to 40 million years throughout the geological record on Earth. The most famous of these bombardments occurred 65 million years ago, when a 10- to 20-kilometer-wide object struck the Yucatán peninsula, spewing a huge cloud of debris into the atmosphere and creating worldwide havoc that apparently exterminated the dinosaurs. Wetherill emphasizes that if the giant planets had not thrown most of the comets into the Oort cloud and had not diverted most of the comets headed for the inner solar system into orbits that keep them far beyond that inner realm, then the rate of bombardment on Earth would have been, and would continue to be, far greater than its actual value. In that case, instead of mocking Californians for their refusal to admit that a giant earthquake, often called the "Big One," will sooner or later destroy much of what they have built, the entire world could await the "Truly Big One"—the impact that would exterminate life on Earth every few hundred thousand years. (Of course, we are hallucinating here; if comets like the "great Shiva," the dinosaur-destroyer that ended the Cretaceous era, had hit the Earth even every million years, then life would not have evolved to the forms we find today.)

Hence, says Wetherill, if we hope to find an Earth-like planet with life, we must find one or more Jupiter-like protectors to keep it safe from the comet cloud of its parent star. To put this another way, finding Jupiter-like planets at Jupiter-like distances would not merely verify the standard model but might also be the key to allowing life to evolve, relatively unscathed, close to the star. As discussed below, Jupiter-like planets close to their stars, like the majority of the recently discovered planets, not only fail to *protect* any Earths but also must have *prevented* any Earths from forming, if the theories of how these planets came into existence has merit. Finally, just to provide our planetary-formation model with an exactness it does not yet deserve, consider what would happen with complete protection against bombardment. If no comets or asteroids had struck the Earth during, say, the past three billion years, life on Earth might have taken quite different paths. After all, the extinction of the dinosaurs cleared the way for us mammals, who spent more than a hundred million years grubbing for whatever the dinosaurs allowed us (not much, unless you favor a shrewlike existence). The evolution of intelligent life may turn out to require a roughly Earth-like planet, along with Jupiter-like planets to protect it—but not too completely!—from the cometary planetesimals left over from the epoch of planet formation.

Don't Forget the Star!

According to the standard model of planet formation, the central parts of the protoplanetary disk create a star while the remainder of the disk becomes colliding planetesimals. Because self-gravitation proves most effective in attracting matter toward the center of the disk, most of the disk's original mass—probably more than 99 percent in most cases—clumps into a single object, first a protostar and then a star. Seen from a nonplanetary, mass-chauvinist perspective, everything outside the star-in-formation amounts to so much chaff. All the objects orbiting our sun, including its trillion comets and trillions upon trillions of meteoroids, have a total mass less than two Jupiter masses: not even 1/5 of 1 percent of the sun's mass. After recognizing a human tendency toward "mass chauvinism," we may admire star formation not only for its effect on the bulk of the matter in a protoplanetary disk but also for the ways that a protostar affects any planets that may be forming around it.

What happens to the protostar, the mass of gas at the center of the protoplanetary disk, that turns it into a star? Self-gravitation squeezes this gas, heating it by compression, until finally the central parts of the protostar grow so hot that nuclear fusion occurs. By astronomers' definitions, this turns the protostar into a star. Because we squeeze gas only rarely in our daily lives, we lack an intuitive feeling for a basic law of nature: Compressing gas to a higher density makes it hotter. The contraction of a protostar, caused by the mutual gravitational attraction of all its particles, raises the temperature of the gas that forms the bulk of the protostar's material. The increase in both the temperature and the density of the gas in the protostar produces an increase in the gas *pressure*, which measures the product of those two quantities. If the protostar's self-gravitation remained constant, its rising internal pressure would soon halt the contraction. In fact, however, this self-gravitation not only grows stronger as the protostar contracts, because all the protostar's parts move closer to all the other parts, but in fact self-gravitational forces increase *more rapidly* than the countervailing effects of an increased pressure. As a result, the protostar continues to grow smaller, denser, and hotter; if the onset of nuclear fusion did not halt the contraction, the protostar could become a black hole without ever being a star.

What a Difference Nuclear Fusion Makes!

The rising temperatures produced when a protostar contracts make the molecules within it collide more and more rapidly, until they all break apart into individual atoms. Further collisions, still more violent because the temperature has risen still further, then break the *atoms* apart into *electrons* and *nuclei*. By this time, the temperature has reached many hundred thousand degrees. The protostar has become densest at its center, most rarefied near its outer boundary, because its self-gravitation tends to pull more material toward its center. As the central temperature passes one million degrees, the protostar approaches the point that marks its transition to stardom. Finally, when the core temperature rises to about 8 million degrees on the absolute scale (about 14 million degrees Fahrenheit), hydrogen nuclei (protons) are moving so rapidly that some of their collisions cause them to fuse. At lower temperatures the protons have smaller velocities, and none of their collisions allow them to overcome their mutual electromagnetic repulsion to meet, kiss, fuse, and lose their existence as protons.

Nuclear fusion turns energy of mass—Einstein's $\mathbf{mc^2}$, the energy that lies locked in every particle with mass $\mathbf{m}$—into new kinetic energy. In stellar interiors, a series of nuclear-fusion reactions changes four protons into one helium nucleus; as this occurs, about 1 percent of the total mass contained in the original four protons disappears, to be replaced by an amount of new energy equal to the amount of "lost" mass times $\mathbf{c^2}$, the square of the speed of light. This new energy appears in the form of more rapid motions of the particles emerging from the fusion reactions and spreads to nearby particles through collisions. Since fusion occurs at the star's center, this moves the new energy outward. Eventually, the new energy spreads all the way to the star's surface, from where the star radiates it into space. By producing energy in its core and spreading that new energy through its entire volume, the star achieves—thanks to the physics of the situation—a long-lasting balance between its tendency to contract, which results from its self-gravitational forces, and its tendency to expand, which arises from the high temperatures in its interior. Neither expanding nor contracting, the star will last for millions or billions of years, as the new kinetic energy released by nuclear fusion in its core slowly but steadily diffuses to the surface.

We should not fall into the trap of imagining that a protostar begins to shine only when nuclear fusion at its center turns it into a star. Any protostar emits radiation at an increasing rate, with the bulk of the emission at progressively shorter wavelengths, as the protostar grows increasingly hotter and denser. As the protostar begins to contract, most of its emission consists of microwaves, but as the contraction continues, its steadily increasing amounts of radiation consist mainly of infrared. Although it cannot match the energy output of a star, a protostar produces significant amounts of infrared radiation during the few tens of millions of years immediately preceding its stellar existence. By the time that it emits large amounts of visible light, the protostar has become a star.

Brown Dwarfs: When a Protostar Fails to Become a Star

Not every protostar becomes a star; some remain protostars indefinitely because they have too little mass to squeeze their innards to the point that nuclear fusion occurs. The characteristic that makes all the difference is the star's *mass*, which determines the amount of the star's self-gravitational forces and thus how high the temperature rises in its interior. Low-mass protostars will never raise their central temperatures to the 14 million degrees Fahrenheit needed for the basic nuclear-fusion process to begin. Higher-mass protostars, in contrast, surpass the 14-million-degree threshold and turn into stars without exception.

Where does the boundary between eventual stardom and eternal protostardom lie? So far as astronomers can calculate, the dividing line in mass between an eternal protostar and an eventual star lies close to 8 percent of the sun's mass, equal to about 80 times the mass of Jupiter. Objects with masses smaller than this will never become stars; instead, they are fated to pass their lives as *brown dwarfs*, a good name, coined by the astronomer Jill Tarter, for a protostar that has no other future. Brown dwarfs radiate in the infrared, and even emit some visible light, but this energy comes from their slow contraction, not from the stellar nuclear fusion that lights our skies. To be precise, we may note that the higher-mass brown dwarfs do engage in some nuclear fusion, involving the relatively rare type of nucleus called deuterium. This does not affect the basic distinction we have drawn, which technically refers to the ability to fuse *protons*, the most common type of nucleus, into other nuclei. Deuterium fusion in the cores of the higher-mass brown dwarfs adds nothing major to the heat that the brown dwarfs produce, whereas proton fusion in protostars sufficiently massive to become stars increases their energy output by a thousandfold, or even more.

How Common Are Brown Dwarfs?

As is true for extrasolar planets, the brown-dwarf-discovery scene has changed rapidly during the past few years, and will almost certainly continue to do so. Astronomers currently feel certain of two brown dwarfs, Gliese 229B and the companion to the star HD 114762. In addition, Michel Mayor and Didier Queloz have recently reported finding six additional brown dwarfs by the Doppler-shift method used to detect the first planet around another star. Thus the number of brown dwarfs now known to exist roughly equals the number of extrasolar planets detected until now.

Gliese 229B gained its fame when a team of astronomers led by Shrinivas Kulkarni of Caltech used the Hubble Space Telescope to *see* the object (Color Plate 15). This brown dwarf orbits Gliese 229A (formerly known as Gliese 229), a relatively dim red star about 18 light years from the solar system. The astronomers found the brown dwarf by making a telescopic survey of the regions immediately surrounding nearby stars, using a small disk to block the light from each star to find, amid the stray starlight, the faint glow from a companion. This method, which we described in the previous chapter, can reveal brown dwarfs, which emit their own light, though probably not planets, which only reflect much smaller amounts of starlight.

Once they had found companion objects, the astronomers observed the objects' spectra with the Palomar Telescope to deduce their surface temperatures. Gliese 229B's spectrum impressed them by showing absorption lines from two types of molecules, methane and water vapor, that are destroyed at high temperatures. The astronomers deduced a temperature of about 700 Celsius (1,250 F) for Gliese 229B's outer layers, cooler by 1,400 degrees Fahrenheit than the outer layers of any star. Infrared observations made from the Mauna Kea Observatory show that Gliese 229B produces 60 times more energy in the form of infrared radiation than in visible light, as we would expect for a cool object (cool, that is, by stellar standards).

The distance of Gliese 229B from Gliese 229A is 44 A.U., approximately equal to the distance between Pluto and the sun. The astronomers who study Gliese 229B have observed no changes in velocity that would arise from motion in orbit; nor could they be expected to have done so, since at a distance of 44 A.U. from the star, the companion takes centuries to complete a single orbit. Nevertheless, the fact that Gliese 229A and B have the same motion through space convinces astronomers that they have found a brown dwarf in orbit around a star. Because astronomers have not observed the motion of either Gliese 229A or Gliese 229B, their conclusions about Gliese 229B's a mass derive from the fact that Gliese 229B seems surely to be a brown dwarf, hence less massive than any star. Current mass estimates range from 10 or 20 times Jupiter's mass all the way up to 60 or 70 Jupiter masses.

The companion to the star HD 114762 remains unglimpsed, but its Doppler-shift signature stands out clearly. To find this object, a multiyear observational program, guided by David Latham of the Harvard Center for Astrophysics, employed a rather antiquated telescope, fitted with relatively old-fashioned equipment for recording stellar spectra, to follow the star much more often—on almost every clear night—than would be possible with a modern telescope, for whose time astronomers compete far more vigorously. The lines in the spectrum of HD 114762 exhibit periodic changes in their frequencies and wavelengths. These variations presumably arise from the Doppler effect; if so, they reflect motions of 600 meters per second in either direction from the star's average velocity along our line of sight.

The fact that the changes in the star's spectrum recur with an 84-day period indicates a companion object with a mass at least 11 times Jupiter's mass. Provided that the astronomers have not been observing an orbital system nearly perpendicular to our line of sight (see page 21), this mass falls well below the minimum mass required for a star, 80 Jupiter masses. HD 114762's companion object thus qualifies as a brown dwarf, discovered by the now-familiar Doppler-shift techniques but never directly observed. Just as they do for the new planets described in chapter 1, the Doppler-shift measurements reveal the eccentricity of the star's orbit, which corresponds to the eccentricity of the much larger orbit of the stellar companion. This eccentricity turns out to be 0.38, much larger than that of any planet's orbit in the solar system, and almost precisely equal to the orbital eccentricity of the object orbiting 70 Virginis.

What Distinguishes Brown Dwarfs from Extralarge Planets?

The low-mass boundary of stardom, which is also the high-mass boundary of brown-dwarfhood, equals 80 times the mass of Jupiter. This implies that some brown dwarfs may not be all that many times more massive than the sun's largest planet, which raises a highly relevant question: Where do we set the dividing lines that distinguish a brown dwarf from a massive planet? We expect that brown dwarfs will usually be more massive than planets, because they consist of masses of gas that remained together as fragmentation occurred within an interstellar cloud, whereas planets had to assemble themselves from collisions among large numbers of planetesimals. Thus brown dwarfs are the low-mass end of the distribution of the objects that might become stars—objects that managed to acquire a large mass of gas and dust all at once, masses often a million times greater than Earth's. In contrast, planets had to "work hard for their masses," competing with other planets to attract the relatively small amounts of mass left over from the disk's central condensation, which became a protostar. Nonetheless, a "super-Jovian planet" that managed to build itself up to 10 or 20 times Jupiter's mass would be much like a brown dwarf with the same mass, even though the planet would have a sizable solid core inside its much larger mass of gas.

One possible way to distinguish brown dwarfs from super-Jovian planets rests on the shapes of the objects' orbits. As we have seen, the standard model of planetary formation produces planetary orbits with only modest eccentricities, especially for the larger planets. Double-star systems, on the other hand, often consist of two stars moving in highly eccentric orbits, implying that when a mass of gas split into two, the two protostars did not move around their center of mass in nearly circular orbits. This difference leads some astronomers to specify the distinction between planets and brown dwarfs in terms of orbital eccentricity: Objects significantly more massive than Jupiter are brown dwarfs if they move in noticeably eccentric orbits, but are planets if their orbits are nearly circular. To other astronomers, this seems to be circular reasoning, which begs the question of *how* a brown dwarf differs from a planet. Does this distinction reside only in the mechanism that formed the objects, so that a ten-Jupiter-mass planet, formed by the accretion of planetesimals, might be identical to a ten-solar-mass brown dwarf, formed from a single subclump within a protostellar cloud? This oversimplifies the matter, because, as we have seen, the planet should have a solid core, and also a different mixture of elements, as the result of its slow aggregation. Furthermore, gravitational interactions between the star and its companion can slowly change the companion's orbit, reducing its usefulness as a marker of how the companion formed.

If we used orbital eccentricity as our sole criterion, we would conclude that the newly discovered companion to 70 Virginis, with a minimum mass of 6.6 Jupiter masses and an orbital eccentricity of 0.4, must be a brown dwarf. Far better to study the potential usefulness of the orbital eccentricity, while recognizing that planets may differ significantly from brown dwarfs with the same mass. From our knowledge of Jupiter, Saturn, and the sun, we expect that the ratio of the abundances of the two lightest elements, hydrogen and helium, to those of all the other elements will be different in a giant planet and a star. For Jupiter and Saturn, this ratio is only one-quarter of its value for the sun. Hence astronomers look forward someday to observing planets and brown dwarfs in order to compare these abundance ratios. Such a comparison will establish whether objects with 5 to 10 Jupiter masses differ in their elemental composition, depending on whether they formed by agglomeration from planetesimals or from a single mass of gas and dust.

Based on currently available data, observations and theory prove nicely congruent if we classify the companions to 70 Virginis and 16 Cygni B as brown dwarfs. Then all giant planets' orbits have near-zero eccentricities, while the orbits of brown dwarfs around their companion stars tend to have eccentricities of 0.2 to 0.6. But beware the theory that fits only a modest array of data! An old and completely valid scientific adage states that since some of the data will prove to be wrong, and exceptions to the "rules" implied by a limited data set will soon emerge, any such theory will soon need revision. Keep an eye out for further discoveries of brown dwarfs and extrasolar planets, checking the eccentricities that go with them, to see how well the distinction based on orbital eccentricity holds up in the face of further knowledge.

The Heartbreak of Double Stars

Astronomers still lack a complete theory (how often this phrase seems appropriate in a book like this!) of the processes that occur as the central mass of gas within a protoplanetary disk finishes its contraction to form a star. They do have statistically complete knowledge of the *results* of the processes, for they know the relative abundances of single, double, triple, quadruple, quintuple, and sextuple star systems. In addition, astronomers know a crucial fact about the dynamics of these systems: *Planets seem most likely to form around single stars.* Most small objects originally orbiting one of the stars in a double or higher-multiple star system will either eventually merge with one of the stars or escape from the system completely. In other words, orbits in non–single star systems have much lower stability than do orbits around a lone star.

At least three possible exceptions exist to this general rule. First, if an object orbits much closer to one of the stars than the distance to the second star, that object's orbit can persist indefinitely: Any such object effectively falls under the gravitational rule of just one of the two or more stars in the system. Second, if two stars in a double-star system differ greatly in their masses, so that their ratio of masses exceeds 25 (this is hardly likely to occur, but it might), objects can have relatively stable orbits if they circle the much more massive star at the same distance that the less massive star has, but always stay 60 degrees ahead of, or behind, the less massive star. This requirement marks two Lagrangian points in the double-star system, named in honor of the French mathematician Joseph-Louis Lagrange, who calculated their properties. Each Lagrangian point marks one corner of two equilateral triangles whose other corners are the two stars. At these Lagrangian points, matter can collect as it moves in orbit, as it has, in the form of numerous asteroids, at the Lagrangian points that the sun and Jupiter create. However, as Lagrange showed, if the ratio of masses falls below 25, the orbits of particles at the Lagrangian points are not stable, and the matter will eventually become part of either the more massive or the less massive object.

A more widespread exception occurs for any orbit *much* larger than the spread of the double- or multiple-star system. In this case, the orbiting object "sees" the system, in gravitational terms, as basically a single object whose mass equals the combined masses of its stars. The star system closest to the sun offers a fine example: Alpha Centauri A and B, both sunlike stars, orbit their center of mass with a separation of 23 A.U., close to the distance between the sun and Uranus. At a distance about 1,000 times greater, the dim star Proxima Centauri orbits both A and B. Alpha Centauri A and B could theoretically have planets moving in Proxima-like orbits. Even though the standard model suggests that planets should not form at such greater distances, the evidence that Proxima did form reminds us that the standard model may be only standard, not universal. (On the other hand, Proxima Centauri might have formed elsewhere and only later have been captured by the gravitational force from Alpha Centauri A and B.)

Although this exception to the rule of nonstable orbits applies to every system of stars, we cannot grow overly excited about it in light of the requirement that the planets move at enormous distances from their stars. For a double-star system such as Alpha Centauri, the planets must have orbits many dozen times larger than the stars' separation. This leaves the planets far out in the boondocks, out-Plutoing Pluto by a factor of ten or more. Our prejudices demand the conclusion that planets so far into the dark of interstellar space, so little warmed or lit by their stars, cannot long detain us in our quest to find and to understand other worlds.

Besides, astronomers have been discovering planets—not planets at hundreds or thousands of A.U. from their stars, but planets so close to stars that they boggle astronomers' minds. We must grasp the nettle now and inquire, Why should we rely on the standard model of planetary formation, when it did not predict any planets, especially giant ones, orbiting as close to their stars as the majority of the newfound worlds?

What Are Those Planets Doing So Close to Their Stars?

When late 1995 and 1996 brought the news of the first planets around other sunlike stars, astronomers found themselves seriously reconsidering their previous insights. Once again nature had apparently surprised them with more than had been dreamt of in their philosophy. Nothing daunted, the astronomical experts have absorbed the new information, have pointed out that a careful reading of some of their previous theoretical work would allow for what was found, and have made new speculations, more fully informed by our new accumulation of information. Along just such paths does science proceed to ever more accurate, though always correctable, generalizations about the cosmos.

The masses of the new planets surprised few astronomers, given the existence of Jupiter, with roughly the same mass, and the fact that the Doppler-shift method preferentially reveals the most massive planets. The far more startling result was that five of the first six planets to be discovered move in orbit at distances much less than Jupiter's distance from the sun, and—most amazing of all—three of these Jupiter-size planets orbit at distances from their stars less than 1/7 of the distance from the sun to *any* of the sun's nine planets, including the innermost one, tiny, dense Mercury. From one perspective, this result likewise was not totally unexpected. The Doppler-shift method of searching for planets favors finding closer-in planets, because these planets move more rapidly in orbit than farther-out ones, and also produce more rapid reflex motions of their stars. From another viewpoint, however, we must remember that relative ease of discovery does not by itself imply the existence of the object to be discovered.

Quite suddenly, however, astronomers have had to confront planetary systems, as convincingly present as they are invisible, that contain Jupiter-like planets whose demonstrated existence so close to their stars poses a series of conundrums. Could a planet like Jupiter ever *form* at these distances? Almost all astronomers agreed that it could not. If not, how did these planets attain their present orbits? In the solar system, planetary orbits remain stable for billions of years, and Jupiter, most astronomers agree, has never been far from its current orbit, certainly nowhere near the Earth's distance from the sun and, most emphatically, not at 1/20 of the Earth-sun distance. In that case, what mechanism exists that could move giant planets from the regions where they form to much smaller distances from their stars? If such a mechanism exists, why do the planets "stop" moving, rather than sidling all the way into a fiery merger with their stars?

What Makes Planets Migrate?

The astronomical world sought answers to these questions from the corps of astronomers who theorize about planetary formation, and who responded by improving their old ideas and creating new ones. As early as 1982, one of the leading planet-formation theorists, Douglas Lin of the University of California, Santa Cruz, had suggested mechanisms that could lead to planetary migration from orbit to orbit. Had Lin's suggestions been widely trumpeted before the discovery of the new planets, the discoveries of 1996 might have borne him on a tidal wave of fame; as it is, he and his fellow theorists must settle for being highly competent astrophysicists, almost unknown to the public but well respected by their peers. The new planets have thoroughly energized Lin and his fellow theorists. Alan Boss of the Carnegie Institution of Washington, one of the leading proponents of the standard model, says, "I think these close-in, 'oddball' planets are great news—if they exist, this gives me more confidence that 'normal' systems [those more like the solar system] can exist as well. Of course, it could be that the solar system is the oddball type, but my feeling is that this will turn out not to be the case."

Let us see how the theorists can explain finding at least four giant planets orbiting their stars at distances less than Mercury's distance from the sun. The solar system contains one object with far more mass than all others: the sun, whose gravitational forces keeps the planets moving in orbit around it. We have seen how planets owe their orbital motion to their birth within a rotating protoplanetary cloud; the orbit that each planet acquired then reflects the balance between the inward pull of the sun's gravity and the tendency of all moving objects to keep moving in a straight line. In a system containing one star and one planet, the planet and star will each orbit around the center of mass essentially forever, without a change in the planet's orbit (which would produce a corresponding change in the star's orbit). If more than one planet exists, then the mutual gravitational forces among them can change their orbits, but our own solar system offers proof that these changes occur on incredibly slow time scales. Half a century ago, a Russian-born psychiatrist named Immanuel Velikovsky published a best-selling book, *Worlds in Collision*, asserting that biblical and historical records implied that Venus and Mars had nearly collided with Earth several times during the past four thousand years. Astronomers judge this a flat impossibility. Velikovsky's vision is reminiscent of the age of bombardment more than four billion years ago, when the planetesimals that had formed the Earth continued to arrive, and a Mars-sized object knocked what would become the moon loose from our newborn planet. To an astronomer, however, a difference in possibility as well as reality exists between what happened four billion years ago, as the solar system ended its formation process, and what is alleged to have happened only a few thousand years ago, at times only one-millionth of the way back to the age of bombardment.

Those who understand the laws of physics and look for possible explanations of planetary migration find the best current cause in the protoplanetary disk, during the eras before nearly all the matter once in the disk had been lost, by merging into planetesimals, evaporating into interstellar space as the result of solar heating and pressure from the solar wind, or spiraling into the sun. In those eras, some four billion years ago, the disk contained a total amount of matter far less than that in the protostar or star that formed at its center; nevertheless, the parts of the disk that lay close to the newly formed planets could have exerted a significant influence on them. Even though the star ruled the roost, gravitationally and otherwise, the modest interaction with the disk could slowly move the planets into different orbits.

But why should the orbits move inward rather than outward,

in view of the fact that the disk presumably lay on both sides of the planets? The answer lies (of course!) in the details of the gravitational interaction between the planet and the disk. When the protoplanets formed, each of them responded to the modest gravitational pull from the nearby parts of the protoplanetary disk. These interactions had opposite effects for the parts of the disk closer to, or farther from, the sun than the protoplanet. The forces from the inner parts of the disk tended to make the planet move more rapidly in its orbit, whereas the forces from the parts of the disk farther from the sun tended to make the planet slow down. A tendency to speed up will produce an outward migration of the planet, but a tendency to slow down will produce a migration inward. (This may seem counter-intuitive, but it is so.) So long as the disk had roughly equal amounts of matter on either side of a protoplanet, the net effect was close to zero, and the planet did not change its orbit.

What kept the amounts of matter from remaining equal in those systems where planetary migration occurred? As the protoplanets and the protostar completed their formation, significant amounts of material in the protoplanetary disk began to move closer to the star. As this occurred, the protoplanet experienced a lack of material closer to the star, in comparison with material farther out (Color Plate 16). This imbalance caused the protoplanet to slow down and move inward, temporarily catching up with the matter closer to the star. As the disk material continued to move inward, so too did the protoplanets, keeping pace with the inward drift of the disk.

Without a protoplanetary disk to produce them, no orbital migrations will occur. If these migrations do occur, they require tens of millions of years for significant changes in planets' orbits. This suggests that sizable migrations will occur only if two conditions are satisfied: The protoplanetary disk must develop an asymmetry, with more disk matter outside the planet's orbit than inside; and this condition must persist for tens of millions of years. If these criteria are met, the planet's orbit will follow the inner parts of the disk inward toward the star. In our own solar system, the protoplanetary disk apparently dissipated within a few million years, and we find the planets orbiting where they formed. Are we the exception, or the rule? So far, the majority of new planets appear to have undergone significant migration inward, but we must recall that the most successful, Doppler-shift method for finding planets preferentially reveals such planets.

What Stops Planetary Migration?

If we can accept that at least in some situations, interactions between protoplanets and the protoplanetary disk can cause planets to migrate inward, we may well ask what stops this migration process and saves the planet from eventually becoming *part* of its star? This problem presents even more difficulty than causing planetary migration to occur. However, we should by now have sufficient confidence in the minds of theoretical astrophysicists to recognize that if they can move them, they can stop them. In collaboration with Peter Bodenheimer and Derek Richardson, Doug Lin has proposed two different mechanisms that could have achieved the desired result.

The more appealing method has the name of tidal interaction, since astronomers use "tidal" to describe any situation in which *different parts of an object feel different amounts of gravitational force*. The best-known tidal interactions are those that raise the tides in the oceans of Earth. As we learned in high school, the amount of gravitational force between any two objects varies in proportion to the product of the objects' masses, divided by the square of the distance between their centers. In theory, the force exerted on a particular object by another object with only a modest mass *might* overwhelm the force from the sun, provided that the low-mass object had a sufficiently small distance in comparison to the distance from the sun. In practice, this never happens. Consider, for example, the gravitational forces that the sun and moon exert on the Earth. The Earth-moon distance equals about 1/400 of the Earth-sun distance, a ratio that favors the moon's gravitational force over the sun's by the square of 400: a factor of 160,000. However, this factor cannot compensate for the fact that the sun has 27 million times the mass of the moon. As a result, the sun exerts 27 million/160,000, or about 167 times more gravitational force on the Earth than the moon does. This is the reason we orbit the sun and not the moon.

But wait! Doesn't the *moon* raise the tides? How can this be, in light of the sun's far greater gravitational force on the Earth? Here we encounter the subtle ways of tidal interactions: The tides arise from *differences* in the amounts of gravitational force exerted on different parts of Earth. For example, the moon attracts each kilogram of matter on the side of the Earth closest to the moon with a greater amount of force than it attracts a kilogram at the Earth's center, because the center lies farther from the moon. Likewise the moon exerts more force on a kilogram of matter at the center of Earth than on a kilogram on the side of Earth facing away from the moon. The differences in force make the Earth bulge both toward and away from the moon. Furthermore (here enters a further subtlety), although both the oceans and the continents respond to these differences in force, the oceans, being liquid, can respond more readily, so the water slides up and down against the continents, producing the ocean tides.

Why do the differences in the amount of gravitational force allow the moon, rather than the sun, to play the crucial role in raising tides? The differences in amount of gravitational force vary in proportion to one over the *cube* of the distance, not to one over the square, as the force itself does. The cube of the sun-moon distance ratio equals, not 160,000, but 400 times that, or 64 million. This number does overbalance the sun-moon mass ratio of 27 million. As a result, the moon has more than twice the tide-raising effect on Earth than the sun does, though the sun can be important. When the sun and the moon are nearly in line with the Earth, close to the times of full moon and new moon, we see especially large tides, called *spring tides*, because (in our forebears' poetic language) the water seems to spring to especially high and low levels as the tides change. Between the spring tides, at times when the sun and the moon are nearly at right angles as seen from Earth (close to first quarter and last quarter in the moon's phases), we encounter *neap tides*, days when the tides seem to take a nap (*nap* and *neap* have the same etymology).

When a planet moves in orbit close to its star, significant tidal interactions occur between the star and the planet. These tidal interactions tend to create a rough match between the planet's orbital period and the star's rotation rate; they also hold the planet in an orbit where such a match occurs. In that case, the planet's orbital period should be measured in days, rather than weeks, months, or years—just what we find for at least five of the planets discovered during 1995 and 1996. Similar tidal interactions occur between a planet and its satellites; they have "locked" our moon's rotation period to its orbital period, so that the moon always keeps the same side toward Earth, and have likewise made the rotation periods of Jupiter's four large moons match their orbital periods.

Hence Lin, Bodenheimer, and Richardson propose that planets stop migrating once tidal interactions with their stars create a match between the planets' rotation rates and orbital periods. The tidal lock has the subtle effect of putting an end to the migration, because any migration farther inward would require a more rapid rotation rate, forbidden by the tidal interaction. This mechanism has its theoretical difficulties, so the astronomical theorists have a second string to their bow, a theory that links the star's magnetic field to the part of the protoplanetary disk that lies close to the star. The stellar magnetic field would change the way that this part of the disk dissipates, so that once the protoplanet reached a distance of about 0.05 A.U. it would experience no significant net tidal influence causing it to move farther inward.

By keeping these two explanations brief, we have avoided both sensory overload and the need to reprogram our brains in case the theories do not hold up subsequent to further investigation, not to mention critiques from other theorists. The crucial point, already grasped, remains the fact that theorists will continue to rise to the challenges that nature offers them, producing improved explanations that in turn will be tested against new data from the cosmos. Plotting the rises and falls of these theories can be left to the experts, who will receive plenty of adverse attention from their competitors—a struggle that gives science its explanatory power. We may now relax a bit by investigating weird planets, both real ones and planets of the imagination.

The Orion Nebula, about 1,600 light years away, is the closest large star-forming region. Amid these clouds of gas and dust, several thousand stars have formed within the past million years; hundreds of thousands more stars may form during the next few million years.

CHAPTER

5

How Strange Can Planets Be?

Humans prefer humanlike issues. In the search for planets, human prejudices lead us naturally toward the question, How much like Earth are other planets likely to be? From an astronomical point of view, the far more interesting way to pose this question, which can shed considerable light on the frequency of Earth-like planets, lies in asking, How varied a range of planets can we expect to find around other stars?

Taking a New Look at Old Worlds

This question cannot receive a full answer until we have acquired a much more comprehensive sample of planetary systems. Nevertheless, working with full consciousness of an incomplete answer, astronomers have striven to calculate and to debate how strange planets can be. Although some broad features of their current answers to this question seem reasonably certain, we would do well to pause to recall the last time that astronomers made entirely reasonable predictions about a set of new worlds.

During the early 1970s, NASA sent twin spacecraft, *Voyager 1* and *Voyager 2*, past Jupiter and Saturn; thanks to careful preplanning of its trajectory, *Voyager 2* received a gravitational-slingshot effect from Jupiter and Saturn that sent it by Uranus (in 1986) and Neptune (in 1989), completing the longest and most successful interplanetary journey to date. The two *Voyager* spacecraft made the first photographic reconnaissance of the moons of the four giant planets, and in particular revealed the surface features of Jupiter's four large satellites, Io, Europa, Ganymede, and Callisto, each at least as large as our moon, and in the case of Ganymede larger than the planet Mercury. The outstanding single conclusion from this survey—a fact that no astronomer had come close to predicting—was that these four worlds show amazing *differences* in their surface features, as well as in other key characteristics. The current reconnaissance of Jupiter and its moons by the *Galileo* spacecraft has emphasized the differences between the satellites' surfaces (Color Plates 17 and 18). Io turns out to be rich in sulfur-spouting, sodium-belching volcanoes, driven by heat produced as tidal forces from Jupiter squeeze the satellite's interior in a continuously changing manner. Europa has a thin coating of ice, which may conceal a moonwide ocean below (Color Plate 19). Ganymede has both light and dark terrain, along with parallel grooves that stretch for hundreds of miles. And Callisto turned out to be the most heavily cratered object found in the solar system, without any of the wide, flat basins that characterize our own moon.

The stunning variety presented by Jupiter's four large satellites finds a resonance in the variegated satellites of Saturn and Uranus. Among these two dozen moons, however, only Titan, Saturn's largest satellite, ranks with Jupiter's big four in size, and Titan's surface remains a mystery, hidden from view by a smog-laden veil of atmosphere. Neptune does have one large satellite, Triton, which resembles the planet Pluto in its overall properties.

To astronomers, the diversity of worlds in the outer solar system came as a complete shock. Of course, no one expected that Jupiter's four large satellites would prove identical, but the data at hand, consisting of how well these moons reflect sunlight and radio waves, together with crude drawings made by planetary observers, had led to a widespread notion that the similarities among these worlds would prove more striking than their differences. More accurately put, astronomers had failed to let their minds grow sufficiently elastic to anticipate how widely divergent in appearance these satellites could be. Could this fate await those who speculate about worlds beyond the solar system?

Before opening this particular can of worms, we should pause a moment to note that in some respects (basically known even before the *Voyager* encounters), the four large worlds around Jupiter, for example, *do* resemble one another. They have roughly similar sizes, masses, and densities, though the two outer ones (Ganymede and Callisto) contain much larger fractions of ice than the two inner ones (Io and Europa). This fact helps to remind us that the question of how different worlds can be cries out for a definition of which sorts of differences most highly engage our attention. And here a fascinating aspect of comparative planetology comes to the fore: Relatively minor differences in the parameters that describe a world (assuming that we can agree on what "minor" means here) may produce major divergences in the fates of those worlds, especially so far as their suitability for life is concerned. No better example of this exists than the comparison of Earth with its two closest planetary neighbors.

Goldilocks and the Three Planets

If we reject as possible sites for life both the planet Mercury, so close to our star that its sunside surface bakes at 600 degrees Fahrenheit, and also our airless moon, the three remaining Earth-like planets—Venus, Earth, and Mars—provide what seem the likeliest places where life might have had a chance to originate. They do so primarily because their surface temperatures from solar heating fall within the range where liquids can exist and because their atmospheric gases contain the most likely elements from which life can be made—carbon, oxygen, nitrogen, and hydrogen. Since the dawn of space exploration, scientists eager to search for life beyond Earth have concentrated on Venus and Mars, but so far without success in finding anything we would call alive. Why is this so? Why have Venus and Mars proven so inhospitable to life, with conditions so unlike those we enjoy on Earth?

Consider, for emphasis, the fact that Venus's bulk properties make it a near twin to Earth in its diameter (95 percent of Earth's), mass (82 percent of Earth's), average density (94 percent of Earth's), and elemental composition (estimated to be nearly identical to Earth's). Yet Venus has an atmosphere one hundred times thicker than ours and made almost entirely of carbon dioxide, a gas that has impressive abilities to trap infrared radiation, holding in heat that would otherwise escape from the planet's surface and lower atmosphere. On Earth, carbon dioxide (CO_2) likewise traps heat to produce what scientists have named, somewhat misleadingly, a *greenhouse effect*. Because Earth's airy blanket contains less than 1/10,000th of the carbon dioxide in Venus's atmosphere, we experience a much smaller greenhouse effect than Venus does. Notice, however, that even a relatively tiny amount of atmospheric carbon dioxide produces a noticeable greenhouse effect on Earth, one that we increase when we release more carbon dioxide by combining oxygen with carbon compounds in combustion. The carbon dioxide in Earth's atmosphere keeps our planet only a few degrees warmer than the temperature that would exist without any CO_2.

On Venus, in contrast, the greenhouse effect produced by carbon dioxide in the atmosphere raises the surface temperature not by a few but rather by about 600 degrees Fahrenheit, so that the planet's surface stifles at temperatures close to 900 degrees! Aside from serving as a reminder of what large quantities of carbon dioxide can do to a planet, Venus's situation raises the crucial issue of what made Venus—at least on its surface—so different from Earth, given their basic resemblances. We might look for the answer in Venus's proximity to the sun: Earth's near-twin orbits the sun at only 72 percent of the Earth-sun distance. This helps to explain the difference, but the basic reason for the apparent dichotomy between conditions on Venus and on Earth lies in an agent one might not immediately suspect: life!

One of life's lesser-sung accomplishments has been the removal of carbon, and thus the potential for making carbon-dioxide molecules, from the Earth's surface. Sea creatures use carbon dioxide dissolved in seawater to make shells and their analogues for microscopic life. Geological processes eventually bury these shells in the ocean floor, slowly make rocks called carbonates (carbon and oxygen combined with other elements), and keep most of these carbonate rocks inert, "locking" the carbon within them. The cycle closes as small fractions of these carbonate rocks, exposed on land by further geological activity, dissolve from weathering, releasing carbon dioxide into the atmosphere; some of the atmospheric carbon dioxide then dissolves in seawater. As a result of this cycle, a modest amount of carbon dioxide exists in the atmosphere at any particular time, with far more in seawater and the rocks formed from sea creatures. If (and we must never try this) we dug up all the carbonate rocks near the Earth's surface, then pulverized and heated them, we would make carbon dioxide by far the most abundant constituent of our atmosphere. In fact, we would thereby liberate about as much carbon dioxide as Venus has in its atmosphere, changing our blue-green planet into the stifling hell we call our neighbor.

To be sure, the ocean would always hold some carbon dioxide whether or not life existed on Earth, and some carbonate rocks would form on the ocean floors, thus reducing the total amount of carbon dioxide in the atmosphere. Nevertheless, the presence of life on Earth has vastly increased our planet's carbon-holding capacity near its surface. Furthermore, life allows the oceans to exist: Without life, carbon-dioxide molecules would be so abundant in the atmosphere that the resulting greenhouse effect would cause the oceans to evaporate. All in all, we tamper with the Earth's greenhouse effect at our peril and have a good example nearby of what a *runaway greenhouse effect* can do: Putting more carbon dioxide into the atmosphere raises the temperature, evaporating more water, which itself increases the greenhouse effect, so the temperature rises still further, and so on and on until we can write *finis* to our planet's happy times.

Mars, like Venus, has an atmosphere composed mainly of carbon dioxide, but instead of being about a hundred times thicker than Earth's, the Martian atmosphere provides less than 1/100 of the surface pressure and density that we find on our planet. Hence the greenhouse effect on Mars, though it does exist, raises the surface temperature by only a few degrees. This may be cause for regret, since Mars is *cold*. At the Martian equator at noon on the warmest day of the year, the temperature may rise almost to Boston standards, but by midnight on the same day, midwinter temperatures at the South Pole would provide the proper comparison, for the thin Martian atmosphere cannot retain much of the daytime heating. Day and night, summer and winter, the average surface temperature on Mars equals about 40 degrees below zero—nearly 100 degrees Fahrenheit below Earth's average temperature. What explains this difference?

The primary cause, as we suspect, arises from Mars's greater distance from the sun, 1.52 A. U. in comparison with the Earth's 1.0. As a result of this distance, each square foot on Mars's surface receives only about half as much solar energy as does a square foot on Earth. A secondary cause of the difference, however, lies in Mars's thin atmosphere, which cannot retain heat during the night. Finally, Earth's atmosphere produces a larger greenhouse effect than Mars does, because it contains water vapor as well as carbon dioxide, both of which trap infrared radiation effectively.

Mars owes its thin atmosphere to its modest gravitation, which in turn arises from its low mass. Because Mars has only 11 percent of Earth's mass, it has done a poorer job in retaining atmospheric gases. In addition, the low temperature on Mars allows carbon dioxide and water to remain frozen in the planet's polar caps (these are mostly frozen carbon dioxide, or dry ice, with a small admixture of water ice) and in the subsurface layers, which, like the permafrost of Alaska and Siberia, contain significant amounts of frozen water. If we someday chose to engage in planetary restructuring, we might arrange for a one-time heating of the polar caps, which could release sufficient carbon dioxide into Mars's atmosphere to produce a greenhouse effect that would allow liquid water to exist on the planet's surface and would also release large amounts of water now locked in the permafrost. Under the present conditions, the atmospheric pressure falls too low for water to exist in a liquid state, so that any ice heated above freezing promptly sublimates into vapor, as frozen carbon dioxide does on Earth. Mars's surface, however, shows channels almost certainly carved by running water, billions of years ago. We may conclude that Mars once had a thicker atmosphere, capable of holding a pressure lid on liquid water, which the planet lost as its gravitational force proved unequal to the effects of evaporation. The planetary restructuring described above could therefore restore long-vanished conditions on Mars.

For now, however, we see the Goldilocks effect in full flower when we contemplate Venus, Earth, and Mars. Venus's closeness to the sun prevented it from having liquid oceans; the lack of liquid water ruled out the possibility of life; and the absence of life, combined with the high temperature, kept so much carbon dioxide in the atmosphere that the greenhouse effect produces 900-degree temperatures. Mars's low gravitation allowed much of its atmosphere to escape; without a significant greenhouse effect, the planet lost most of its remaining carbon dioxide and water to its polar caps and permafrost. Only Earth has maintained conditions that allow liquid water to persist—and liquid water may be essential for life (see chapter 8).

What Makes Planets Different from One Another?

We have seen that the sun's planets and their larger satellites exhibit startling, significant differences in their sizes, appearances, compositions, temperatures, and atmospheres. Given that we inevitably survey new worlds with one eye on their potential habitability, we naturally encounter the crucial question, What makes these worlds so different?

Until 1995 astronomers thought they had a straightforward answer to this question, one that would apply throughout the cosmos and that arises from the processes that formed a planetary system. The crucial influence on a planet was the *location* at which it formed. According to the standard model of planet formation, the sun's four giant planets owe their large masses to their formation at large distances from the sun, where more matter existed in the protoplanetary disk and lower temperatures allowed the protoplanets to retain hydrogen and helium more easily. The four terrestrial planets, in contrast, forming in locations with less material, never had a chance to retain much of the two abundant gases. The recent announcement of planets around Lalande 21185 (see page 133) with masses similar to Jupiter's, moving in orbits whose sizes resemble those of the sun's giant planets, provides support for this analysis—provided the planets turn out to be real!

Red Dwarfs, Brown Dwarfs, and Planets

The standard model for planetary formation helps to distinguish an extremely massive planet from an extremely small star, and thus to answer the question of just what characteristics distinguish a planet from a star. Here astronomers impose a twofold test, which examines what celestial objects *do* and also how they *formed*. Stars liberate kinetic energy in their interiors through the process called nuclear fusion; planets engage in no such activity. Furthermore (so astronomers believe), stars formed through a "top-down" process, during which a single mass of gas contracted under its self-gravitational force, whereas planets agglomerated from countless smaller objects through a "bottom-up" formation scheme. The distinction based on function strikes the eye, since nuclear fusion allows stars to generate light on their own, but the difference in formation has equal importance when we survey the cosmos to ask which objects deserve the name of planet. If by *planet* we mean an object something like our own world, we must look not only to the present state of the object but also to the process that produced it.

These twin distinctions play a key role when we consider the class of objects that seem to straddle the star-planet boundary: the brown dwarfs described in the previous chapter. Brown dwarfs are objects with too little mass to be stars yet are denied the title of planet because they presumably formed in a top-down manner, rather than accreting from collisions among planetesimals. Why do astronomers think that the formation of brown dwarfs resembled the processes that made a star rather than those that produced a planet? They draw this conclusion largely from computer models of the formation process, which imply that the least massive clump of gas and dust that results from the fragmentation of an interstellar cloud will have a mass about ten times the mass of Jupiter. This suggests (though the suggestion may be one we adopt because it makes for more convenient packages in our minds) that objects with more than about ten Jupiter masses are brown dwarfs, while those with less than this mass are planets.

What does the term *planet* mean when we draw this distinction? We have drawn a bright line between brown dwarfs and planets on the basis of how they formed, but if we were to look at the objects themselves, what differences would we see between a 12-Jupiter-mass brown dwarf made from a contracting clump of gas and dust, and an 8-Jupiter-mass planet that formed from colliding planetesimals, themselves made within a protoplanetary disk of matter?

Not too much. The 8-Jupiter-mass planet would have a solid core of perhaps 25 to 100 Earth masses (0.08 to 0.3 Jupiter masses), the heart of the planet that formed from dirty ice within the protoplanetary disk and allowed the massive planet to acquire large amounts of hydrogen and helium. The brown dwarf would consist mainly of hydrogen and helium all the way through, with no solid core. Both objects would consist mostly of hydrogen and helium, the lightest and most abundant elements. Computer models imply that both types of objects should have a diameter close to Jupiter's, despite their greater masses, because their greater self-gravitation compresses them more effectively.

If a giant planet or a brown dwarf orbits extremely close to a star, at distances comparable to 0.05 A.U., as is true for four of the first eight extrasolar planets, the star will noticeably affect the planet's overall configuration. First of all, heat from the star will warm the planet, making its gaseous layers expand. We may expect that Jupiter-like planets close to their stars should be "fluffier" than similarly made planets orbiting at much larger distances. This should make the close-in planets comparatively easier to see—were it not for the greater difficulty of avoiding the far greater glow from the star (see page 58). Second, the planet will develop a significant tidal bulge, so that its diameter from pole to pole becomes significantly less than its diameter across the equator. The tidal bulge arises from the differences in the amount of gravitational force exerted by the star on the planet's near side, center, and far side. Similar effects cause the Earth's oceans to bulge slightly toward and away from the moon, but the tidal bulge that occurs when a planet orbits within 0.1 A.U. of its star puts Earth's tidal changes to shame. Even though we have never seen the planet around 51 Pegasi or the inner planet around 55 Cancri, we may conclude that they must be more flattened than Jupiter or Saturn, with equatorial diameters at least 20 percent greater than their diameters from pole to pole.

Strangest So Far: Planets around Pulsars

One happy aspect that the planetary discoveries of 1995 and 1996 offered to astronomers was their relief at having "real planets," rather than the only planets definitively established to exist beyond the solar system—the three planets that orbit the pulsar PSR 1257+12.

Planets around pulsars! What could be odder, astronomers

COLOR PLATE 9 The Very Large Array (VLA) of radio telescopes has a Y-shaped configuration that can spread its 27 antennas across more than 20 miles of dry plains in central New Mexico; this aerial view shows the antennas in a crowded arrangement. Astronomers employ these 27 antennas as a much-valued interferometer to obtain high-resolution images of radio-emitting sources.

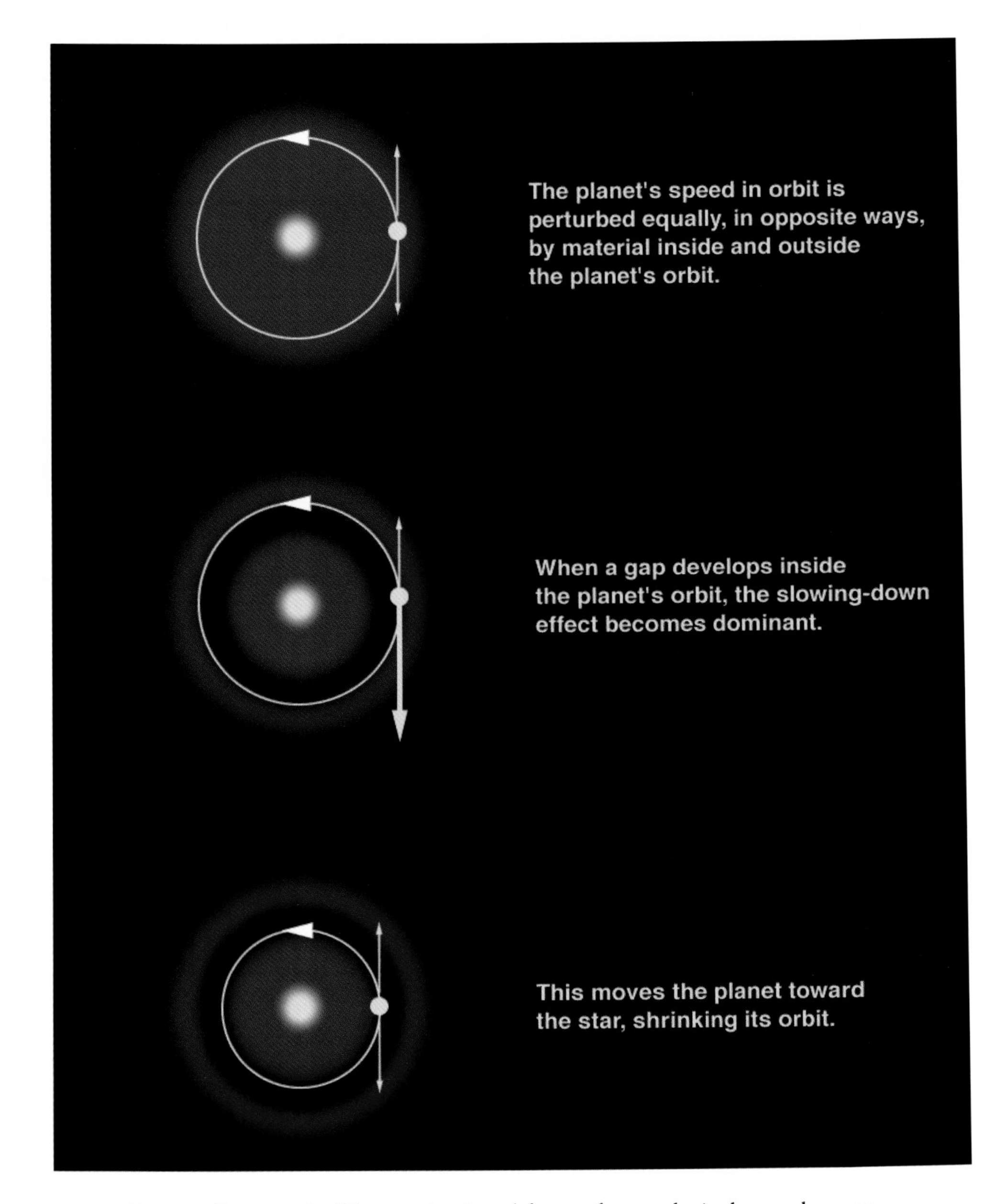

COLOR PLATE 16 The gravitational forces from relatively nearby matter in the protoplanetary disk can change a planet's orbit. Matter farther from the star tends to slow the planet in orbit and make it move inward, while matter closer to the star tends to speed up the planet, causing it to move outward. If the amount of matter close to the planet on either side of its orbit is roughly equal, then the planet will not move. However, if significantly greater amounts of matter lie just outside the planet's orbit than just inside it, the planet will move inward.

asked themselves, upon hearing the news in 1992. Pulsars arise from *neutron stars*, stars that have collapsed at the ends of their nuclear-fusing lives to create an amazing object with more mass than the sun packed into a sphere no larger than Chicago. Still more amazing from a planetary perspective, the collapse that forms a neutron star also initiates a titanic explosion, a *supernova*, which blows the star's outer layers into space at speeds of thousands of kilometers per second. Conventional wisdom held that a supernova explosion would surely destroy any planets that might exist around the aging, pre-supernova star. Yet there were the data, explicable only on the basis of planets in orbit. Furthermore, these planets had *Earth-like* masses; they still rank as by far the least massive objects discovered outside the solar system, with masses less than 1/100 of the mass of the least massive objects so far discovered—indeed, of any objects that *could* be discovered with astronomers' present techniques—around ordinary, sunlike stars such as 51 Pegasi and 70 Virginis.

How could astronomers find objects with such tiny masses, orbiting a pulsar whose distance (1,600 light years) far exceeds the distances to any of the sunlike stars now known to have planets? The answer lies in the fact that pulsars create highly accurate timing devices, which astronomers can observe with impressive precision. This occurs because of the mechanism that drives pulsars: the rotation of neutron stars.

All neutron stars are born from the collapsed cores of stars. Astronomers believe that these cores have some rotation. We can observe the rotation of stellar surfaces, not only our sun's (which takes about a month to rotate once) but also of other stars, because the rotation broadens the stars' emission and absorption lines, spreading them over wider bands of frequency and wavelength as the Doppler effect increases the frequency of the light from the star's approaching side and decreases the frequency of starlight from the receding side. The inner parts of stars almost certainly rotate more rapidly than the outer regions, but their rotation periods are still measured in days. However, when these central regions collapse to form a neutron star, their rotation rates increase dramatically. The phenomenon, called *conservation of angular momentum*, which allows high divers and ice skaters to spin more quickly by contracting their sizes, implies that an object's rate of spin will increase in inverse proportion to the *square* of its size. If, for example, a diver tucks her body to half its extended length, she will spin four times more rapidly. When a star's core "dives" inward, it decreases its size not by a factor of two but by a factor of a thousand or so: The neutron star produced by the collapse has a diameter of only a dozen miles, instead of the tens of thousands of miles characteristic of a star's energy-producing interior. This raises the rotation rate by several million times, so that the newborn neutron star spins many times per second.

The rapidly spinning collapsed core—now called a neutron star because it consists almost entirely of neutrons—also possesses an intense magnetic field, which likewise increased in strength as the result of the collapse that formed the neutron star. As the neutron star spins, this magnetic field sweeps into motion the nearby debris left over from the collapse, mainly elementary particles called protons and electrons. As the magnetic field exerts its electromagnetic force on the protons and electrons, it actually accelerates some of them until they reach speeds close to the speed of light. At this point, the fast-moving particles produce radio waves (and sometimes infrared, visible-light, and x rays as well). They do so because radio waves, along with some higher-frequency electromagnetic radiation such as infrared and visible light, appear whenever charged particles move at nearly the speed of light in the presence of intense magnetic fields and change either their speed or their direction.

Physicists call this process *synchrotron emission*, because they first analyzed it when they accelerated charged particles to nearly the speed of light in instruments they had named synchrotrons. Since the synchrotron-emission process occurs whenever charged particles reach near-light velocities in the presence of magnetic fields, astronomers might well choose a more appropriate name, such as magneto-violent emission, to refer to these situations, but tradition keeps the term *synchrotron emission* in astronomical daily use.

The few hundred miles surrounding a neutron star possess in plenty all the elements needed to produce synchrotron emission—magnetic fields, charged particles moving at nearly the speed of light, and changes in the particles' speed or direction. The forces from the magnetic field that accelerate the particles to near-light velocities also make them continuously change their directions of motion. The neutron star's surroundings therefore produce intense radio-wave emission. Synchrotron emission has the characteristic, not shared by most familiar sources of radiation, that the radiation emerges in beamed form, rather than in equal amounts in all directions. If some regions close to the rapidly spinning neutron star emit especially large amounts of radio waves, these regions create a sort of cosmic lighthouse, whose intense beam of emission rotates along with the neutron star (Color Plate 20). Astronomers' computer models exhibit just this behavior, with "hot spots" producing strong emission in beams that arise from kinks in the magnetic field or from similar situations. If our line of sight toward a neutron star happens to coincide with a direction that the beam sweeps over, we shall detect an especially strong pulse of radio waves each time that the neutron star's rotation carries the beam past us. We shall then observe a *pulsar*, the immediate environment of a neutron star.

Pulsars are born with rapid rotation, but they slow down as they age, because the act of emitting radio waves and other types of electromagnetic radiation takes kinetic energy from the spinning neutron star. Of course, in recording the times at which the pulses arrive from a particular pulsar, astronomers must take account of the Earth's rotation, the Earth's motion around the sun, and the sun's motion around the Milky Way, all of which constantly change the distance from the pulsar to us and thus affect the intervals between successive pulsars; however, over the years astronomers have learned to do this quite well. Pulsars now provided the most accurate clocks we find in nature, more precise, for example, than the Earth's rotation (which undergoes small jerks or hiccups from time to time). Some pulsars occasionally show glitches in their pulsation periods, which apparently arise from "starquakes" that slightly change the size of the neutron star that powers them. Between these glitches, though, pulsars provide clocks that are accurate to the eighth decimal place or better.

During the 1980s astronomers found a new class of pulsars: millisecond pulsars, with pulse periods measured in thousandths of a second. From the slow-down history of neutron stars, we might feel sure that millisecond pulsars must be the youngest, but instead they rank among the oldest pulsars! According to the analysis of the pulsar experts, each millisecond pulsar once formed part of a double-star system, in which each star orbited the system's center of mass. The explosion that produced the pulsar must have changed these orbits as the supernova shot a large part of its mass into space but still left a binary system of which the neutron star formed one member and a relatively normal star, the other. As the normal star aged, it swelled to become a red giant, as all stars do, and eventually began to lose mass from its outer layers. Some of this material, falling onto the neutron star's surface from a preferential direction mandated by the stars' orbital motions, caused a "spin-up" in the rotation rate. Over thousands of years, this spin-up could have raised the rate from perhaps once every few seconds, for an old pulsar, to the many hundreds of times per second typical of millisecond pulsars.

By carefully timing the pulses from millisecond pulsars, astronomers who specialize in these observations found the evidence that in many cases they had found a pulsar moving in orbit with a companion object or objects. The pulses would arrive a bit sooner, then a bit later, then a bit sooner again than would be expected from the average interval between pulses. These changes in pulse-arrival times have a direct analogy to the changes in *velocity* of an orbiting object, which produce the Doppler shifts we have already examined in some detail. With millisecond pulsars, however, astronomers can directly measure the changes in the *distances* that the radio pulses must travel to reach us. Like the velocity changes, these variations in distance arise from the object's motion in orbit, and, like them as well, they can show only that part of the motion that lies along our line of sight. Careful analysis of the pulse-arrival times will reveal both the orbital period (the time for a complete cycle to recur) and the velocity with which the pulsar moves (since the pulse-arrival times reveal the changes in distance, and these changes equal the speed in orbit times the relevant time interval). As in the case of sunlike stars, the period and velocity provide crucial information about the mass and distance of the object producing the observed motions.

Early in the 1990s Alex Wolszczan, a radio astronomer at Pennsylvania State University, found a pulsar with companions that had masses far too small to be stars. Wolszczan employed the Arecibo radio telescope, the world's largest radio dish and one equipped with superbly accurate timing devices to register the arrivals of pulses, to observe the millisecond pulsar PSR 1257+12. (The letters in this designation stand for "PulSating Radio source," and the numbers identify the pulsar's position with respect to the celestial equator.) Located in the constellation Virgo, about 1,600 light years from Earth, PSR 1257+12 emits pulses 161 times per second (the actual interval between pulses equals 0.0062185319388187 second, a tribute to astronomers' abilities in timing pulses; we may casually round their result to 6.2 milliseconds). The interval between pulse arrivals steadily increases, because the pulsar has a recession velocity along our line of sight. Its distance from us therefore grows continuously greater, so each successive pulse takes a bit longer to reach us. Wolszczan's observations revealed that *superimposed on that steady increase are even smaller, variable changes in the intervals between successive pulses*. These intervals ran first a bit longer than the expected value, then reversed their behavior to be timed at shorter-than-expected values, then ran long again, and so on in cyclical repetition. The deviations from the expected values in the pulse-arrival times were far too small to arise from perturbations by a companion *star*, nor did they show the characteristic sine-curve shape expected if a single object were orbiting the neutron star.

Wolszczan collaborated with Dale Frail, an astronomer

working at the Very Large Array of radio telescopes in New Mexico, to establish the exact position of the pulsar, necessary to avoid unwanted contributions to the timing measurements. Their analysis of the pulse-arrival times, published as 1993 began, showed that PSR 1257+12 has not one but two objects in orbit; a year later Wolszczan demonstrated the presence of a third orbiting object (Table 5.1). Each of these three objects tugs the neutron star, so the pulse-arrival times show the superimposition of three cycles, each with its own orbital period. Nevertheless, thanks to accurate observations and well-programmed computers, the astronomers could "deconvolve"the record of pulse-arrival times to find each of the three cycles. In fact, there are now hints of a fourth planet around this pulsar, whose orbital characteristics and mass have not yet been accurately determined. In velocity terms, the accuracy attained by Wolszczan and Frail for the first three planets around PSR 1257+12 amounts to measuring the speed of a faraway object, not to a precision of 10 meters per second, or even 1 meter per second, but 1 *millimeter* per second—3,000 times better than the finest Doppler-shift measurements in visible light!

Because of its extreme precision, the timing of the pulses from PSR 1257+12 could and did reveal planets with less mass than the Earth. These are in fact the first three extrasolar planets to be discovered: Earth-like in their masses and orbital sizes, revealed by their tiny gravitational effects on a marvelously well-timed pulsar. Yet few astronomers, and not many in the wider public, grow excited about these planets, at whose appearance we can only guess. What seems to be the problem here?

Only the site where the planets had been found. The pulsar planets are "illegitimate," born under the "wrong" circumstances. These planets "have no business" appearing around a pulsar, the scene of an explosion so astronomically recent, and so violent, that the conventional picture of planet formation seems to have no applicability, for the star that exploded must have destroyed any planets that had formed with it long ago. In addition, the explosion would even more certainly have snuffed out any life

Table 5.1

Planets Found around PSR 1257+12, with Earth and Jupiter for Comparison

Planet Designation	*Planet's Minimum Mass (in units of Earth's mass)*	*Average Planet-Pulsar Distance (in A.U.)*	*Orbital Period (days)*	*Orbital Eccentricity (see note to Table 1.1)*
A	0.015	0.19	25.34	0.00
B	3.4	0.36	66.54	0.018
C	2.8	0.47	98.22	0.026
Earth	1.0	1.0*	365.24	0.02
Jupiter	318	5.2*	4332.6	0.05

*Distance to sun

forms we can easily imagine, so the pulsar planets offered small encouragement to those who see the exploration of the universe as a chance to find life. As things turn out, however, the planets around PSR 1257+12 can teach several important lessons to those who search for more familiar sorts of planets around more familiar stars. We may summarize these lessons with four words: thoroughness, imagination, coherence, and pancakes.

First of all, the pulsar planets show that nature often reveals what you don't expect to find—if you remember to keep looking. As the physicist Edward Kolb puts it, "the most difficult thing to discover is something for which you are not searching." The fact that astronomers, like other scientists, know that they must produce results in order to win their peers' respect tends to focus their plans, but the price for this focus can be the loss of awareness to unexpected results. Alex Wolszczan remained aware of the possibility of finding low-mass objects in orbit around millisecond pulsars, and he and Dale Frail recognized that the complex changes in the pulse-arrival times from PSR 1257+12 could occur when two or more objects exerted gravitational forces on the neutron star. As a result, they found the pulsar planets in a situation where other astronomers might well have missed them.

Second, the cosmos has a tendency to surpass our imagination. Intimately connected with point one, this fact, by its nature, remains difficult to keep in mind. Some unexpected discoveries occur from the mental combination of two or more results, none of them by itself surprising. Others, such as finding planets around pulsars—or the discovery of a Jupiter-like planet orbiting a sunlike star at only 0.05 A.U.—straightforwardly break the mind-mold. In hindsight, the earlier mental rigidity had far less justification than had been believed: easy to say, once sudden illumination has appeared.

Third, the coagulation of matter in situations related to the formation of planets may prove surprisingly simple, in light of the fact that at least one pulsar has planets. When astronomers failed to make mental provision for the possibility of planets around pulsars, they had good reasons. Even the "old" pulsars represented by the millisecond pulsars have ages measured in millions of years, not hundreds of millions or billions, reckoning from the time that the star exploded. Furthermore, the violence of the explosion not only destroyed any planets then in orbit but also ejected large amounts of matter from the immediate vicinity of the dying star, hardly a prime setting in which to imagine new planets forming. But form they did, and we find the *only* certain case of extrasolar planets that are Earth-like in their masses in orbit around a pulsar. Astronomers know nothing about the physical appearance of these planets (see Color Plate 20 for an artist's conception), though they strongly suspect they must consist mainly of heavier elements such as silicon, oxygen, carbon, and aluminum; the lightest elements, hydrogen and helium, should have been expelled into space by the supernova explosion, leaving no chance for the postexplosion surroundings to form gas-giant planets.

The pulsar planets testify both to the validity of the standard model and to the cosmos's ability to make planets. Since astronomers still trust the calculations showing that a supernova would blow apart any Earth-like planets within a few A.U., the pulsar planets must have somehow formed, or reformed, from the debris that did remain in the vicinity of the star that exploded, and whose core has produced PSR 1257+12. If planets can be made amid big-city violence, so to speak, perhaps they can easily be made anywhere: The immediate vicinity of an ordinary star-in-formation seems to offer conditions far more favorable to planet formation than do the surroundings of a recently exploded star. Not only does a protostar's environment possess more raw material than the surroundings of a supernova, but it also has far longer spans of time for planetesimals to form and for planets to accrete from them in relative serenity.

Fourth, as to pancakes: The pulsar planets tend to confirm many of the basic components of the standard model of planet formation. Like the sun's planets, the pulsar planets all orbit in approximately the same plane (otherwise, their effects on the pulse-arrival times could not be so easily disentangled). This implies that planets formed as condensations within a flattened disk of material, rotating around the massive object at its center, both in the solar system and around PSR 1257+12. When we combine this fact with the discussion in the paragraph above, the concept that matter close to a massive object naturally forms a disk, and clumps within that disk produce planetlike objects, seems ever more attractive.

Let us not wander too far from the Popelike wisdom we embraced on page 5. On the one hand, PSR 1257+12 and the solar system arguably givers us two, not just one, planetary systems that each contain several planets. "Two" marks the start of statistics rather than scientific stamp-collecting, but we would feel much happier with a dozen or so multiplanet systems to compare and contrast. Happily, that result seems to lie just over the millennial horizon.

How Many More Pulsars Have Planets?

By now, astronomers have discovered nearly a thousand pulsars, of which more than 30 belong to the millisecond class, with periods shorter than 0.01 second. Of all these pulsars, various astronomers have observed about 300 with a timing accuracy sufficient to reveal planets with relatively long orbital periods—half a year or more. Only about 10 of the millisecond pulsars have received the much greater scrutiny needed to find planets like the ones around PSR 1257+12. In the summer of 1996 the radio astronomers Zaven Arzoumanian, Kriten Joshi, Fred Rasio, and Steve Thorsett announced the discovery of three planets around one of these pulsars, PSR 1828-11. The planets have orbital periods of 0.68, 1.35, and 2.71 years; they orbit at distances of 0.93, 1.32, and 2.1 A.U.; and their masses are 3, 12, and 8 times the Earth's mass. Thus these planets' orbits resemble the orbits of the sun's inner planets even more closely than the orbits of the planets around PSR 1257+12. In addition, the same astronomers have apparently found an object with about 10 Jupiter masses—more than 3,000 Earth masses—in orbit around the pulsar PSR 1620-26. This may well be a brown dwarf, provided that we can imagine a brown dwarf forming around the remnant core of an exploded star.

Thus astronomers have now found planets around two or three pulsars, with no planets detected around a far greater number of pulsars. To be sure, astronomers don't much expect to find planets around the younger (nonmillisecond) pulsars, on the ground that their debris has not had time to agglomerate into planetlike objects. Among the millisecond pulsars, those with companion stars still in orbit may well have conditions that inhibit, or prohibit, the formation of smaller condensations, just as astronomers think may be the case for double- and multiple-star systems (see page 94). But explanations are cheap, while evidence talks. The discovery of planets around PRS 1828-11 rules out the possibility that PSR 1257+12 is a rogue pulsar system.

Anything Stranger Still?

Now that we have loosened our mental bonds with a look at planets around pulsars, can we soar still farther from Earth-like realms, to consider worlds still odder than those made from reconstituted, postexplosion debris? Let's try a few categories of strange and see how they fly. First, planets might exist that drift through space on their own, not gravitationally bound to any star or more massive object. Such a planet might have formed around a star, only later finding its freedom, perhaps as the result of a close encounter with another object in its original planetary system. Or it might somehow have originated on its own, as a clump of matter within an interstellar cloud that grew to planet size without being part of a protoplanetary disk around a protostar.

The first possibility leaves the planet "strange" only in the lack of light and heat from its star. If a planet formed in the "usual" way, it might undergo any number of changes, including loss of its atmosphere through evaporation or freezing, without significantly changing its basic composition and size. In contrast, a planet that formed on its own might well baffle the theorist and boggle the viewer. The planet could, for example, consist largely of carbon atoms, locked into long-chain molecules, as the result of its formation from enormous numbers of interstellar dust grains. Intriguingly enough, the surface of a planet like that would be a marvelous place for life to originate—if only one could supply a solvent and the means of keeping that solvent liquid (see page 39).

Why Can't a Planet Be Square?

Among the various speculations we can make about planets that may orbit other stars, one conclusion has the merit of being almost certainly correct: Any naturally occurring object large enough to be called a planet—that is, any object larger than a few hundred miles across—will have a spheroidal shape, thanks to gravity. Any object's self-gravitation tends to make it spherical, the shape for which the average distance of each part of the object from every other part falls to the lowest possible value. At least three other influences affect this tendency toward sphericity. First, a solid object resists deformation: If it forms with a nonspherical shape, it will maintain this shape unless its self-gravitation produces stresses large enough to overcome the strength of the object's material. This occurs for objects larger than a few hundred miles in diameter; in contrast, smaller objects can maintain highly elongated and contoured shapes indefinitely. Second, an object's rotation makes it bulge in the equatorial direction and contract along its polar axis. For this reason, Earth has an equatorial diameter 1/300 longer than its polar diameter. For the giant planets Jupiter and Saturn, this excess rises toward 1/10, making the planets spheroidal rather than spherical. The third influence that can make a planet nonspherical arises from tidal distortion, that is, from another object's gravitational force (see page 101). As we saw, tidal influences from the nearby star surely distort close-in planets such as the one around 51 Pegasi. Like the effect of a planet's rotation, however, tidal distortion produces a spheroidal planet.

In sum, though planets may vary widely in their sizes, masses, compositions, and distances from their stars, every single planet at least as large as Jupiter's Galilean satellites should have neither sharp corners nor protrusions that extend dozens of miles above their surfaces. Although mountains on Earth can rise a few miles high, temporarily resisting the gravitational forces that tend to make our planet a sphere, they could not be much higher without succumbing to gravity. A solid planet twice as massive as Earth should have a lower limit to the height of its mountains, while a less massive planet might have ten-mile-high mountain ranges. We should note, though, that a five-mile-high mountain rising above the surface of an eight-thousand-mile-wide planet represents only a modest deviation from smoothness—and that no solid planet roughly the size of Earth will have higher mountains.

Of course, nothing verifies a conclusion such as this one about planets that may circle other stars as well as actual observation of the objects in question. It is time to consider the different ways in which we might search for extrasolar planets, always aware that planets with different characteristics may reveal themselves in preferentially different ways.

The two stars in the double-star system closest to the sun, Alpha Centauri A and B, produce a twin set of "diffraction spikes" in the telescope. The light from each of these stars outshines by several billion times the reflected light from any planets that might move in orbit around the center of mass of the double-star system.

CHAPTER

6

Seven Ways to Find Planets Around Other Stars

Fate arranged that the Doppler-shift technique would be the first method to succeed in finding planets around sunlike stars. But in their searches for extrasolar planetary systems, astronomers have more arrows in their quivers, each of which holds promise. Collectively, these methods provide a highly significant complementarity. Some of them work best for planets close to their stars, others for planets at greater distances. Some work well at all distances, others only for the closest stars. Some yield results quickly, others only over decades. And some offer the chance to make repeated observations of other planetary systems, while others are one-shot affairs.

As we have seen, direct observation of extrasolar planets does not rank among the top methods, though its day will surely come. Let us recapitulate and examine seven promising methods, keeping an eye on their complementarity and thus on the chances that they can reveal planets of different types and at different distances from their stars. Our tour will take us through searches based on astrometry, planetary transits, Doppler shifts, pulsar timing, gravitational microlensing, interferometry, and the search for extraterrestrial intelligence. Two of these have already met with success, as outlined in chapter 1 (Doppler shifts) and chapter 5 (pulsar timing); the other five have yet to reveal planets, though astrometry seems poised on the brink (see below). To one of these methods, optical and infrared interferometry for the direct imaging of planets, we shall devote the final chapter of this book, since the new millennium seems likely to bring it to fruition and thus to yield our first direct observations of worlds around other stars.

Astrometry: Measuring the Stars in Their Courses

One of the oldest aspects of studying the sky consists of *astrometry*, the measurement of stellar positions—that is, the determination of their locations with respect to one another on what astronomers still call the celestial sphere. More than 1,500 years ago, Claudius Ptolemy introduced his concept of the universe as consisting of nested, transparent spherical shells, each one centered on the Earth and carrying an important part of the cosmos as it rotated. The innermost shell carried the moon, while the outermost one bore the "fixed stars." Today the term *celestial sphere* refers to our view through what Ptolemy would have considered *all* of the shells surrounding the Earth. Since all celestial objects seem to participate in a rotation that carries them across the skies, we can still imagine that they are dots on an imaginary spherical shell that turns around the Earth.

In fact, it is the Earth that rotates, not the heavens. Our planet's rotation produces the *appearance* of celestial rotation that seems to carry the sun, moon, planets, and stars across the sky. Careful observations show that the sun, moon, and planets move against the background of "fixed stars," which Ptolemy imagined to have identical distances from Earth. (These movements are what led Ptolemy to his model of nested crystalline spheres.) Even though the stars have widely different distances from the solar systems, astronomers often find the celestial-sphere concept a useful one in describing how to locate objects on the sky. But the notion of a single outer shell that carries the "fixed stars" loses accuracy over time. Astronomers have also discovered that the stars' relative positions are not "fixed"; instead, over long intervals of time, stars move with respect to one another, as if they were sliding around the celestial sphere rather than remaining glued to a single position.

The individual motion of a star has the name *proper motion*, where "proper" has its older meaning of "belonging to a particular object." To detect these proper motions, astronomers record the relative positions of stars in a specific direction at many moments in time, spread over a decade or more. They must, of course, allow for, and subtract, the Earth's own yearly motion around the sun. When they do so, they find that most of the stars in any one field of view are so distant that they do remain fixed in position with respect to one another, but a minority of stars show changes in their positions against the backdrop of much more distant stars.

What produces a star's proper motion? All the stars in the Milky Way, including our own, are in constant movement around the galactic center, so proper motions arise from a combination of our own motion and that of the star under observation. In other words, what we observe is the star's motion with respect to the sun's; if these motions were absolutely identical, astronomers would observe zero proper motion for that star. In fact, however, stellar orbits around the galactic center are not identical, so most stars have some motion with respect to the sun. These relative motions typically amount to 5–50 kilometers per second, not much when we consider that all the stars are moving around the galactic center at speeds of about 250 kilometers per second, but enough to produce an observable amount of proper motion.

Proper motions are measured in terms of how rapidly a star appears to change its position on the celestial sphere, where locations are specified in terms of angles. For stars within a dozen light years of the sun, these changes typically amount to a few seconds of arc per year. As we look to more distant stars, we can expect the average proper motion to decline in direct proportion to the distance, because the amount of a star's proper motion depends on its velocity with respect to the solar system divided by its distance. The proper-motion record holder among the stars surrounding the solar system has the name of Barnard's Star, because the American astronomer Edward E. Barnard discovered its large proper motion nearly a century ago. Barnard's Star, six light years from the sun, ranks just behind the Alpha Centauri system (4.4 light years) on the list of nearby stars.

As astronomers extended and improved on Barnard's observations, some of them found a tantalizing phenomenon. Instead of occurring in a perfectly straight line, as would be expected for any star observed for less than one part in a million of a complete orbit around the galactic center, Barnard's Star's proper motion apparently deviated slightly, first to one side and then to the other of the general straight line, as if a less massive object were tugging it in different directions as time passed (Color Plate 21). Further observations suggested that two, possibly three, objects were moving in orbits with different periods, combining or opposing their gravitational pulls as time passed. For a time, Barnard's Star seemed the best candidate for a star with planets—not a sunlike star, to be sure, since Barnard's Star is a dim, reddish dwarf with less than one-fifth the sun's mass and only one one-hundredth of the sun's luminosity, but a star all the same. The planets around Barnard's Star, with masses comparable to Jupiter's and Saturn's, orbiting at distances comparable to Jupiter's distance from the sun and with orbital periods measured in tens and twenties of years, might well receive little light and heat, but they would have been true planets all the same.

"Would have been" notes the fact that the observations proved erroneous. Astrometry attempts to measure tiny angles, and many effects—most noticeably those introduced by changes in the atmospheric transmission of light—can produce tiny errors that can be mistaken for real changes in a star's position. The future of astrometry lies in space, where observations can be made free from the distortions introduced by Earth's atmosphere. During the mid-1990s a European spacecraft called *Hipparcos* has made millions of measurements of stellar positions from its orbit around the Earth, attaining a precision beyond that achievable by any Earth-based observatory. Of course, decades must pass before this type of observing program reveals planets, because the proper-motion approach works best for planetary orbits with periods significantly greater than one year. The effects of shorter-period orbits tend to be "lost in the noise" produced by the Earth's annual motion around the sun. Furthermore, a star moving through space at a particular velocity will exhibit a proper motion that decreases with the star's distance from the observer. Since the chance of finding *changes* in the proper motion—changes that presumably arise from planets in orbit—decreases as the amount of proper motion diminishes, this means that astrometry has a chance of revealing extrasolar planets only for the stars closest to the solar system.

The difficulty of measuring changes in stars' proper motions should not be taken to imply that Earth-based astrometric observations simply cannot reveal the existence of planets. At the Allegheny Observatory in Pittsburgh (hardly the best astronomical site on Earth, but in a location once endowed by well-to-do donors), the astronomer George Gatewood refurbished a large refracting telescope with a technologically advanced detector system to search for deviations from straight lines in the motions of nearby stars. In addition to his own observing program, begun in 1988, Gatewood has analyzed nearly six decades of photographic plates taken by his predecessors. The program proved that the "planets" around Barnard's Star do not exist.

In June 1996 Gatewood announced that his project *had* found planets around a star, Lalande 21185, which appears just after Barnard's Star on the list of stars closest to the sun. Lalande 21185 is a dim red dwarf, much like Barnard's Star and 8 light years from the solar system. Despite their rank among the sun's nearest neighbors, both Barnard's Star and Lalande 21185 have such low intrinsic luminosities that they cannot be seen with our unaided eyes. From his mass of observations, Gatewood has deduced the existence of at least two planets around Lalande 21185, one with a mass at least 0.9 times Jupiter's mass and a

5.8-year orbital period, the other with somewhat less mass, moving in a larger orbit with a 30-year orbital period. Gatewood's data also suggest the existence of a third planet at even larger distances from the star. His results, if confirmed, would make Lalande 21185's planets the first multiplanet system found around a true star, with an impressive resemblance between these planets' orbits to those of Jupiter and Saturn, the sun's two largest planets, whose orbital periods are 12 and 30 years, respectively.

However, many astronomers, keenly cognizant of the Barnard's Star episode, remain dubious, awaiting further data before they state a firm opinion about these planets. Geoff Marcy and Paul Butler have studied this system without finding planets, though at the star-planet distances reported by Gatewood, the planets would move relatively slowly, so that Doppler-shift observations would have only a marginal ability to detect them. One month after Gatewood's announcement, Marcy and Butler announced their first multiplanet system, the two objects around 55 Cancri (see Table 1.1). For now, we may take a conservative approach and state that highly suggestive, though not yet conclusive, evidence exists for planets around the fifth-closest star to the sun (counting the three stars in the Alpha Centauri system separately). Should Lalande 21185's planets be confirmed, the implication would be clear: Planetary systems are abundant in our corner of the Milky Way, for we must look only to the fifth star outward to find another system.

The Line-Up: When a Planet Transits Its Star

Because the sun's innermost planets, Mercury and Venus, orbit closer to the sun than we do, they can pass directly between the sun and the Earth. This occurs less often than one might think, because the orbits of these two planets are tilted slightly with respect to the Earth's orbit. The passage of Mercury or Venus directly between the Earth and the sun produces a *transit*, causing Mercury or Venus to appear as a small black dot as the planet crosses the solar disk, blocking a tiny fraction of the sun's light. Transits once provided astronomers with important information, useful in determining the distance to the sun; astronomers would travel to far-off lands to observe the transit from widely separated locations and thus to triangulate their lines of sight toward the sun.

When astronomers consider the configurations that might carry a planet directly between ourselves and its star, some speak of *transits*, while others, using an older and technically incorrect term, talk of *occultations*. In this book, I shall use only the word *transit*; in any case, what counts is the thing rather than the word. Suppose, for example, we lived on a planet around one of the tens of thousands of stars that lie within a hundred light years of the solar system. In that case, the sun would appear as a star of mediocre brightness. If we made a careful survey of thousands of stars, seeking changes in their *apparent brightnesses*, most locations would never see a significant change in the sun's apparent brightness. However, if our location just happened to be in line with the plane that contains Jupiter's orbit around the sun, then every 11.86 years, when Jupiter happened to pass directly between the sun and ourselves, we *would* notice a dip in the sun's brightness. How large a dip? Observing from a distance of many light years, we could reasonably assign the same distance to Jupiter and the sun, since the sun-Jupiter distance is less than 1/10,000 of a light year. In that case, because the sun has a diameter 9.8 times Jupiter's, Jupiter would appear to cover just $1/(9.8)^2 = 1/96$ of the sun's surface area, so the sun's apparent brightness would diminish by just over 1 percent.

Good monitoring equipment would therefore reveal a decline in the sun's apparent brightness as Jupiter passed in front of it. How long would this decrease last? Jupiter moves in orbit at about 13 kilometers per second, and the sun has a diameter of about 1.4 million kilometers. Jupiter therefore takes about 100,000 seconds—about 28 hours—to transit across the sun, as seen by any observer at a large distance (here *large* means a distance much larger than the sun-Jupiter distance). If other stars have Jupiter-sized planets, the detection of the transits of these planets will reveal them, and their orbital periods as well, provided that our line of sight happens to coincide with the planets' orbits around their stars and provided that we have the determination to monitor the stars' brightnesses for several decades, waiting to observe successive dips in brightness.

What about finding an *Earth* by the transit method? Here we wait not 11.86 years but only 1 year between successive transits. On the other hand, the sun's diameter exceeds the Earth's by not 9.8 but 109 times, so the Earth covers only $1/(109)^2 = 1/11,881$ of the sun's surface. To discover an Earth by observing its transit across a sunlike star, we therefore require the ability to find a change in brightness by about one part in 12,000! Since Earth moves in orbit at 30 kilometers per second, its transit across the 1.4-million-kilometer-wide sun, and thus the tiny dip in the sun's brightness, would last for 47,000 seconds (about 13 hours). Note that neither the amount of the dip in brightness nor the duration of this transit depends on the distance from the observer to the planetary system, so that in theory we could detect transits just as well at distances of thousands of light years as at a dozen light years. In practice, however, fainter objects are more difficult to observe accurately than brighter ones, especially when we search for a dip in brightness by a tiny fraction of a percent. We would therefore do well to concentrate on the closer sunlike stars, which have the greater apparent brightnesses, with any transit-observation program we may create.

Using the transit method to search for planets moving in Jupiter-like orbits seems a task for long-lived masochists. A single dip can never be convincing, because stars have their own internal readjustments, which often cause a modest fluctuation in brightness for a few hours or a few days. The crucial evidence for a planet would appear in the cyclical recurrence of such a dimming event, repeating at time intervals that directly reveal the planet's orbital period. For planets with orbits comparable to Jupiter's, that means waiting at least a decade, and—for greater accuracy—a couple of decades, to observe at least three successive dips in brightness. For a planet like Saturn, with a diameter 85 percent of Jupiter's and an orbital period of 29.5 years, this technique would require noting three dips of about 0.7 percent in brightness over a century.

But this hardly counts as an objection to the transit method, for no one plans to study stellar brightnesses for a century to search for planets in Saturn-like orbits. (Of course, if we had started forty years ago . . . but then the task hardly seemed worthwhile.) Instead, the plans for using transits to detect planets center on searches for planets in orbits similar to the Earth's, and they aim to create a telescope devoted to transit searches, placed into orbit around the Earth, above the atmosphere that blurs all observations. This automated system, originally named *FRESIP*—for *FR*equency of *E*arth-*S*ized *I*nner *P*lanets—has now been renamed the Kepler Mission in honor of the great astronomer Johannes Kepler. Still nowhere near the funding stage, the Kepler Mission would use a wide-angle telescope to monitor the brightnesses of about 50,000 stars. By observing from space, astronomers can reasonably hope to detect dips in brightness as small as 1 part in 12,000, or even less, so that the Kepler Mission would have the ability to detect planets at least the size of Earth. Of all the stars that have planetary systems, about one of a hundred should—purely by chance—have its planets' orbital planes aligned almost directly along our line of sight. In that case, the transit of a planet would slightly diminish the starlight for a few hours. Computers would register this fact and would direct the telescope to study such objects even more carefully. By extending the Kepler Mission survey over several years, astronomers could hope to observe several successive dips for a planet in an Earth-like orbit. A second dip, a few months or a year or two after the first, would be highly suggestive, and a third, with a time interval between dips equal to the first such interval, would be almost conclusive. With the orbital period known, one can easily calculate the size of the planet's orbit, and the amount by which the starlight decreases during a transit provides a good estimate of the planet's size.

In 1996 astronomers at Villanova University reported that their observations of a double-star system called CM Draconis imply that a planet is moving in orbit around the two stars. CM Draconis, about 50 light years from the solar system, contains two dim stars, quite close to one another, each of which has about half the sun's mass and only about 6 percent of the sun's luminosity. We know that our line of sight toward CM Draconis coincides with the plane in which the two stars orbit their common center of mass, because we observe periodic eclipses of each star by the other, which reveal that their orbital period equals only 1.27 days. The orbits of any planets circling the center of mass of the double-star system are likely to lie in the same plane, assuming that the planets formed from the protoplanetary disk of the double-protostar system. On June 1, 1996, an automated telescope and detector registered a 7-percent decline in the light from this orbital pair, at a time when the stars were not themselves lined up to produce an eclipse as we see them. Other observers had recorded two similar drops in brightness 735 days earlier, in May 1994. This provides suggestive, though hardly definitive, evidence for one or more planets orbiting the double-star system with a period equal to 735 days (or possibly one-half, one-third, one-quarter, or some similar fraction of 735 days—the system has not been monitored with sufficient consistency for all such brightness drops to have been noticed). Astronomers will continue to monitor the CM Draconis system to test the hypothesis that they have found the first planet to be detected by the transit method.

The transit method of searching for planets has the advantage that, unlike the astrometric method, it can find planets around stars out to distances of hundreds of light years. Furthermore, because astronomers well understand the geometry that determines the chance of an orbital alignment that produces transits, even negative results from such a search could yield important data on the frequency of planets at least as large as Earth. Along with the gravitational-microlensing method discussed later in this chapter, the transit method offers the only means of detecting Earth-like planets (not counting the possibility of contact with other civilizations!). However, a space-borne planetary-transit observing project has yet to gain widespread support within the astronomical community; we should be willing to exhale long before it becomes reality.

Which Planets Produce Detectable Doppler Shifts?

The most successful method to date for finding planets around other stars is the Doppler-shift method, described in chapter 1, which relies on accurate observations of a star's velocity along our line of sight to detect periodic variations, presumably caused by a planet's gravitational pull as it orbits the star. The changes in stars' velocities produce Doppler shifts, changes in the frequencies and wavelengths of the features in the stars' spectra. These Doppler shifts arise from the Doppler effect, the fact that motion toward or away from an observer will change the frequencies and wavelengths of light that the observer receives.

How large will those changes be for a particular star-planet situation? They are easy to calculate, thanks in part to the fact that in any system in which a less massive and a more massive object orbit their common center of mass, the product of mass times velocity will be the same for the two objects. For example, as the Earth orbits the sun, its gravitational force pulls on the sun, just as the sun pulls on the Earth. As a result (ignoring the other planets for the moment), the Earth and the sun both orbit the center of mass of the Earth-sun system, which lies on the line joining sun and Earth, with the distances of the two objects from the center of mass in inverse proportion to the objects' masses. Since the sun has about 330,000 times the Earth's mass, the center of mass lies 330,000 times farther from the center of the Earth than from the center of the sun. This puts the center of mass deep inside the sun, only a small fraction of the sun's radius away from its center—but the sun makes its own yearly orbit around this center of mass.

Because the product of mass and velocity is the same for the sun and the Earth, their relative velocities vary in *inverse* proportion to their masses. The sun's velocity in response to the Earth's gravitational force is just 1/330,000 of the Earth's velocity in response to the sun's gravity. Since the Earth has an orbital velocity of about 30 kilometers per second, the *sun*'s orbital velocity amounts to 30/330,000 kilometers per second, or about 10 *centimeters* per second. If any other star with the sun's mass has a planet with one Earth mass, moving in orbit at the same distance as the Earth-sun distance, then that star also will have an orbital velocity of 10 centimeters per second. To find our twin, "all" that we need do would be to measure cyclical changes in the star's velocity by this amount!

The difficulty lies in using the Doppler effect to measure velocity changes of only 10 centimeters per second. Since light travels at nearly 300,000 kilometers per second, a velocity of 10 centimeters per second represents about one-third of one-billionth of the speed of light and produces a Doppler shift by the same fractional amount in the frequencies and wavelengths of any spectral features. Amazingly advanced though modern techniques may be, they cannot yet measure Doppler shifts anywhere nearly this small. As described in chapter 1, the best we can do—and mighty fine it is, too—is to measure Doppler shifts of about one part in a million, which corresponds to velocities just one-millionth the speed of light, or about 3 meters per second. Not bad, considering that this amounts to finding the speed of a brisk walk in comparison to light speed! You can't find an Earth with this accuracy, but you can hope to find a Jupiter.

Why is this? Jupiter moves more slowly in its orbit than Earth does, because Jupiter orbits farther from the sun. The velocity of any planet orbiting a sunlike star (that is, a star with approximately the same mass as the sun) equals 30 kilometers per second divided by the square root of the planet's distance, measured in A.U. Note that the velocity does not depend on the planet's mass: Even though the star exerts more gravitational force on a more massive planet, the planet's resistance to acceleration by a force (called its "inertia" in older texts) likewise increases in proportion to the planet's mass, so it's a wash. A good thing too, or else the Earth and the moon could hardly orbit the sun at the same rate. Since Jupiter is 5.2 A.U. from the sun, its orbital velocity equals 30 divided by the square root of 5.2, or 13, kilometers per second. If Jupiter orbited the sun at the Earth's distance, Jupiter would move at 30 kilometers per second, and the sun's velocity would vary by plus or minus 30 meters per second in response to Jupiter's gravitational force.

Color Plate 22 shows the current situation for the Doppler-shift method. The graph presents the limit of detectability, based on the fact that 3 meters per second represents the smallest velocity shifts that can be detected at the present time. Above this diagonal line, we may expect planets to reveal themselves by producing measurable Doppler shifts, but we cannot now find any planets that may exist below the line. This limit of detectability rises toward higher masses for greater distances from the star, because a planet of any particular mass will produce smaller motions in its star as its distance from the star increases. Note that Jupiter lies above the limit of detectability, while Saturn, with 30 percent of Jupiter's mass, lies just below the line. Within the next decade, therefore, we may expect to learn that astronomers have found Jupiter-like planets—provided that these planets exist!—not only at distances up to 2 A.U. from their stars, but at 4, 8, or even 10 A.U. However, when we recall that planets at these distances will take from 7 to 30 years to orbit their stars, we can see that they will not reveal themselves on time scales measured in months or even a few years. Only when astronomers see at least one complete cycle in the velocity changes deduced from a star's spectrum can they reasonably conclude that they have found a planet, and one complete cycle lasts exactly as long as one complete orbit.

The Doppler-shift method of searching for other planets now works very well, so long as the planets have relatively large masses and orbit relatively close to their stars. This method inevitably favors *finding* just those planets with relatively large masses and small distances from their stars, but especially those with large masses. Indeed, this effect appears in the first six planets discovered with this technique: All six have Jupiter-like masses, and five of the six orbit their stars at distances much less than the distance between the Earth and the sun.

Pulsar Timing: What Hope for the Future?

In the previous chapter, we saw that the incredible accuracy in timing the arrival of radio pulses has allowed astronomers to deduce the existence of three planets around the pulsar PSR 1257+12, and apparently three more around the pulsar 1828-11. This method therefore ranks as one of two—indeed, the first—actually to reveal extrasolar planets. Astronomers hope soon to have sufficiently precise observations of several dozen pulsars to assess whether this pulsar may most accurately be seen as a fluke, or whether planets exist around other pulsars too.

However, so far as familiar planets go, we cannot hope for success from the timing method, simply because ordinary stars have nothing to time. If only every star emitted pulses with the repetitive exactness of millisecond pulsars! In that case, we could use pulse-arrival timing to discover any planets with Earth-like or greater masses around any individual star we can observe, not only in the Milky Way but in nearby galaxies as well. This would be too easy, so we must count ourselves lucky that the cosmos provides greater challenges. Note, however, that a civilization similar to our own that wanted to signal its existence could create a radio beacon on its home planet, emitting signals as precisely spaced in time as a pulsar's radio pulses. In that case, another civilization at our own level of technological sophistication could discover this planet moving in orbit, even though it might have a mass much smaller than Earth's.

Gravitational Microlensing: When Planets Bend Light Rays

By far the most counterintuitive method of finding planets around other stars in the Milky Way relies on the phenomenon of *gravitational lensing*, the bending and focusing of light rays by gravitational forces. In 1916 Albert Einstein published his theory of general relativity, which he had labored to bring to completion while Europe lost itself in the First World War (Einstein himself had received considerable opprobrium in Germany as the war began for circulating a manifesto against war as a conflict-resolving mechanism; it received three signatures in addition to his own.) Einstein's general-relativity theory made a startling pronouncement: *Gravity bends space*. Light rays, which we think of as traveling in straight lines (as they do through empty space) therefore deviate from straight-line trajectories when they pass close to a massive object such as a star. The object warps space by the largest amount in its immediate vicinity, and by progressively lesser amounts at increasing distances from the object.

Three years after its publication, with the war over and scientists back at their familiar tasks, Einstein's theory received triumphant vindication at the total solar eclipse of May 29, 1919, when British-led scientific expeditions to Brazil and Principe Island photographed the stars visible near the edge of the moon-covered sun during totality. Home again in Britain, the astronomers carefully compared these photographs with photographs taken of the same region of the sky, but at another season of the year, when the sun was nowhere in the vicinity. The sun seems to move with respect to the stars during the course of a year. As the Earth moves along its orbit, our line of sight to the sun, extended outward, passes successively through all the points around the circle on the sky that astronomers call the "ecliptic." The comparison of photographs revealed that the stars' relative positions at the time of the eclipse differed from those measured at other seasons. Furthermore, the effect was in just the direction, and had just the size, that Einstein's theory predicted: The sun's gravity, bending light rays toward it, made the stars' positions seem to spread outward. The eclipse observations showed that Einstein was right to plus or minus 50 percent; modern observations using the same principles have verified Einstein's predication to within 1-percent accuracy. Einstein awoke one day (November 7, 1919) to find himself the most famous scientist in the world. On the spot, he had to invent the persona that would allow him to function as a scientist while enjoying this fame, and he apparently found that his own personality did the trick quite well. Meanwhile, the world of physics, which had already recognized Einstein's genius since the publication of his theory of special relativity in 1905, had to struggle to understand the primary lesson of general-relativity theory: *Gravity bends space, and bent space "tells" light and objects how to move.*

Fascinating though that struggle can be (see the book by Kip Thorne listed in the Further Reading section for a good introduction), it lies beyond the scope of this work—except for the fact that Einstein's theory leads naturally to the concept of gravitational lensing. If objects with significant amounts of mass bend light rays (and they do), we can imagine some situations whose geometrical arrangement has the result of spreading out light rays, and others in which the gravitational forces tend to focus light rays, producing an increase in an object's apparent brightness. The amount of these effects will depend on the size of the object producing the light, on the size and mass of the object that focuses it, and on the distance between those two objects and the distance from the focusing object to the observer. This amount can be modest, but in particular situations an impressive focusing can occur, one that causes an increase by ten or a hundred times in the object's apparent brightness (Color Plate 23). Astrophysicists use the term *gravitational lensing* to describe situations in which they can see a curve of light, typically a "gravitational arc," resulting from the gravitational bending of light. They describe situations in which they can observe only the changes in an object's apparent brightness produced by gravitational bending as *gravitational microlensing.*

Consider how gravitational microlensing can lead to planet detection. We live on a planet orbiting a star in the Milky Way: The sun and its attendant planets move in a nearly circular trajectory around the center of our galaxy, taking about 240 million years per orbit. Most of the other 300 billion or so stars in the Milky Way have similar motions, likewise moving in circular orbits at greater or lesser distances from the galactic center. As a result, the Milky Way's disklike shape contains an entire galactic ballet, as each star not only circles the center but also bobs up and down slightly, rising a few hundred light years above and below the plane of the galactic disk several times during each orbit. If we examine a distant star at just the time when another, somewhat closer star happens to come between ourselves and the distant star, then the intervening star will produce a gravitational-lens effect, temporarily brightening the light from the farther star. And if the intervening star has a planet in orbit around it, that planet's gravitational force can also produce a lensing effect, far more modest than that from the star, but nonetheless detectable. For our purposes, what counts is the fact that *gravitational microlensing can reveal planets all through the Milky Way!*

In some ways, gravitational microlensing resembles the transit method: Neither depends fundamentally on the distance to the object under observation, and both deal with events that last for several hours. The most favorable lineup for gravitational microlensing occurs when the object producing the lensing lies just halfway between the observer and the source whose light is bent by gravity, but other distances do not diminish the effect much. Of course, in both cases it helps to have a brighter rather than a fainter source to observe. Unlike the transit method, however, where astronomers may be seeking to detect a change in brightness by one part in 12,000, the gravitational microlensing technique deals with increases in brightness by a factor of somewhere between a few percent and (in extremely favorable cases) many dozen.

On the other hand, the geometrical requirement than an object pass almost directly between ourselves and a star makes gravitational microlensing extremely unlikely, much more so than the transit of a planet across the face of its star. This would make the microlensing method only a curiosity—had astronomers not learned how to overcome a statistically unfavorable situation with improved observational techniques. Starting a few years ago, teams of American and European astronomers began projects to use gravitational microlensing to search for objects generally called *machos*. The word stands for a *ma*ssive *c*ompact *h*alo *o*bject, a slightly misleading name since "massive" in this context means "far more massive than an elementary particle, perhaps much less massive than Earth," and "compact" means "not diffuse—a familiar type of object." The "halo" in question, the outer reaches of the Milky Way galaxy, provides a natural place to search—not for planets, but for the "dark matter" that apparently dominates the universe. Dark matter exerts gravitational forces, which allow astronomers to deduce its existence and to measure its average density, but the nature of dark matter remains almost completely undetermined. Microlensing observations drew their original driving force from a search for this dark matter, but within a short time, astronomers have come to realize that these observational techniques also provide an excellent means—with important caveats—of searching for extrasolar planets.

Because the outer regions of the Milky Way seemed the best place to search for machos, the macho teams set up shop in Australia, where they could easily observe stars in the Large Magellanic Cloud, one of the Milky Way's two sizable satellite galaxies, which astronomers refer to as the LMC. A macho anywhere along the line of sight to one of the stars in the LMC would produce a gravitational microlensing that briefly increased the star's apparent brightness. By observing two different colors of light, the astronomers could verify the effect, since only microlensing would affect both colors in the same proportion; by timing the duration of the event and measuring the increase in brightness, they could accurately estimate the mass of the macho.

Only the dearth of near perfect lineups stood in the way, and astronomers solved this problem by observing many million stars each night! Not one by one, of course; the astronomers developed sophisticated camera systems with millions of picture elements, capable of recording the apparent brightnesses of millions of stars and sending their results into a computer that would crank out

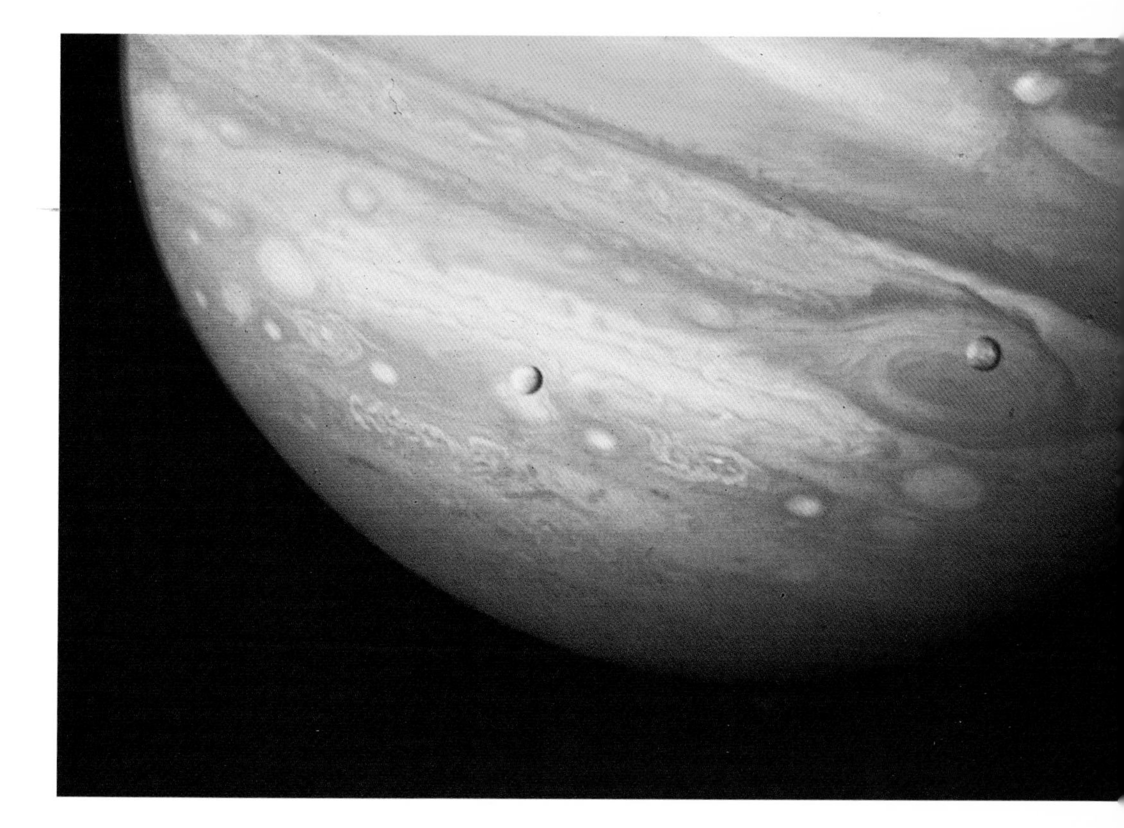

COLOR PLATE 17 Jupiter, the sun's largest and most massive planet, has 11 times the Earth's diameter and 318 times its mass. This view of Jupiter, taken by the *Voyager 1* spacecraft in 1979, shows Jupiter's "Great Red Spot," a semipermanent cyclone area twice as wide as the Earth, with Jupiter's innermost large satellite, Io, directly in front of it. Europa, another of Jupiter's large moons, can be seen to the left of Io.

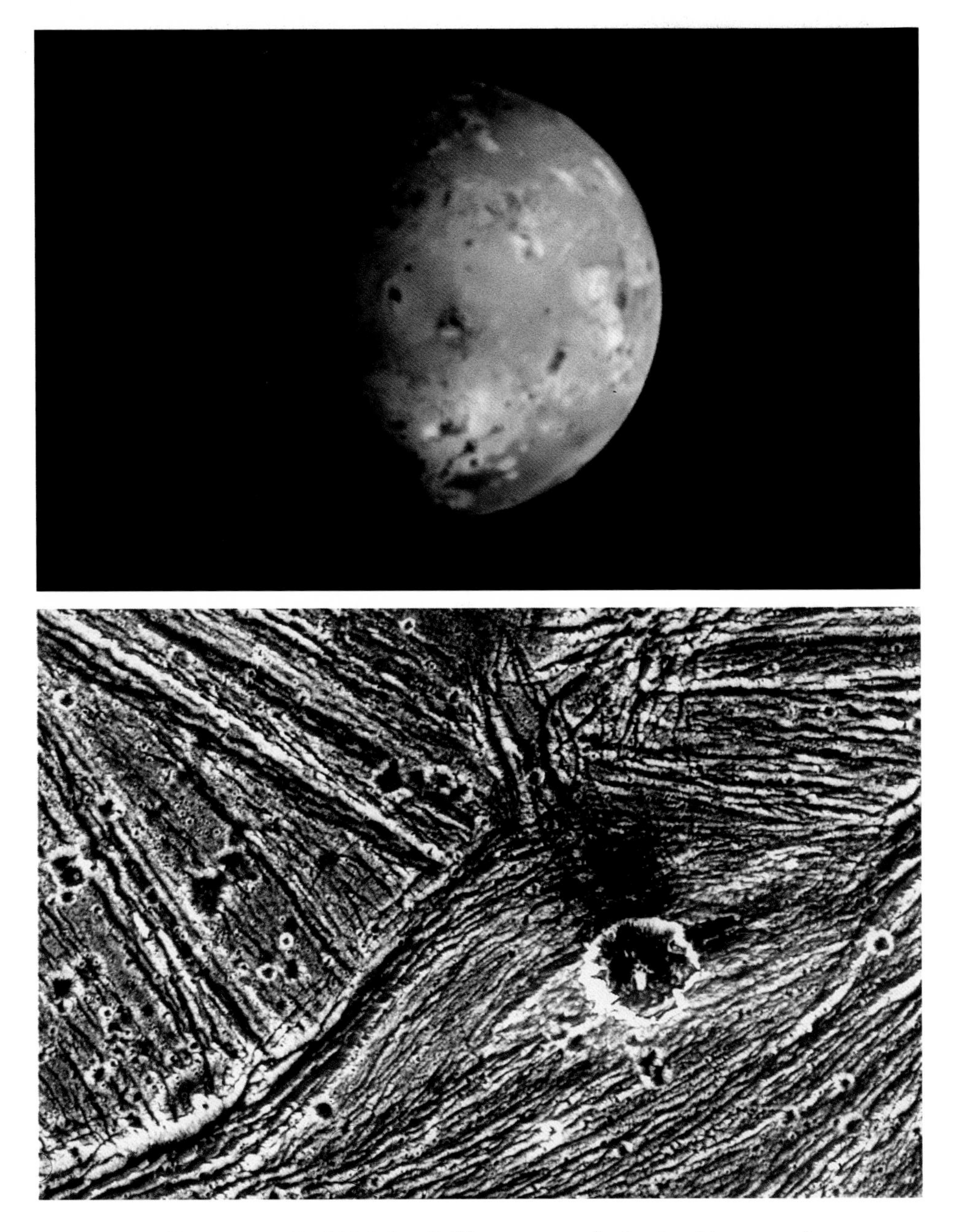

COLOR PLATE 18 In 1996 the *Galileo* spacecraft obtained images of Jupiter's moons Io (top) and Ganymede (bottom), which demonstrate the wide variety of surfaces on Jupiter's four large satellites. Io, about as large as our own moon, has active volcanoes that spew sulfur-rich compounds to produce the variegated colors on Io's surface. Ganymede, the largest moon in the solar system (a bit larger than the planet Mercury), has a surface with few color contrasts. Ganymede's rugged rills and valleys imply that as the satellite formed, its crust cracked and subsided, perhaps as the result of the freezing and remelting of ice below the surface.

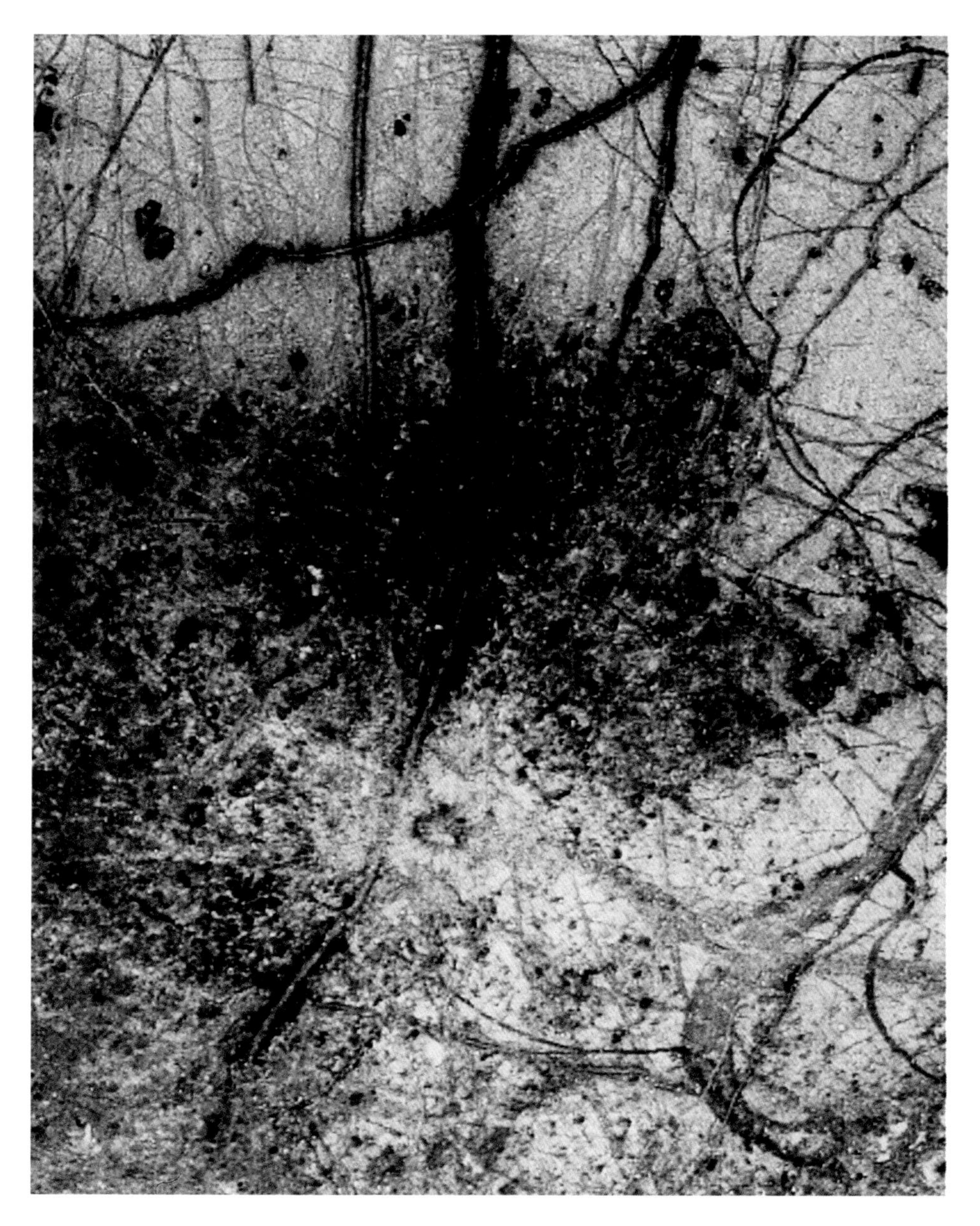

COLOR PLATE 19 This *Galileo* image of Jupiter's large moon Europa looks nothing like Io or Ganymede. Instead, Europa's surface shows long, jagged lines that appear to be cracks in an icy crust. This crust may cover an ocean of liquid, or a global liquid slush, warmed by the moon's internal heating, making Europa a fine prospect for extraterrestrial life.

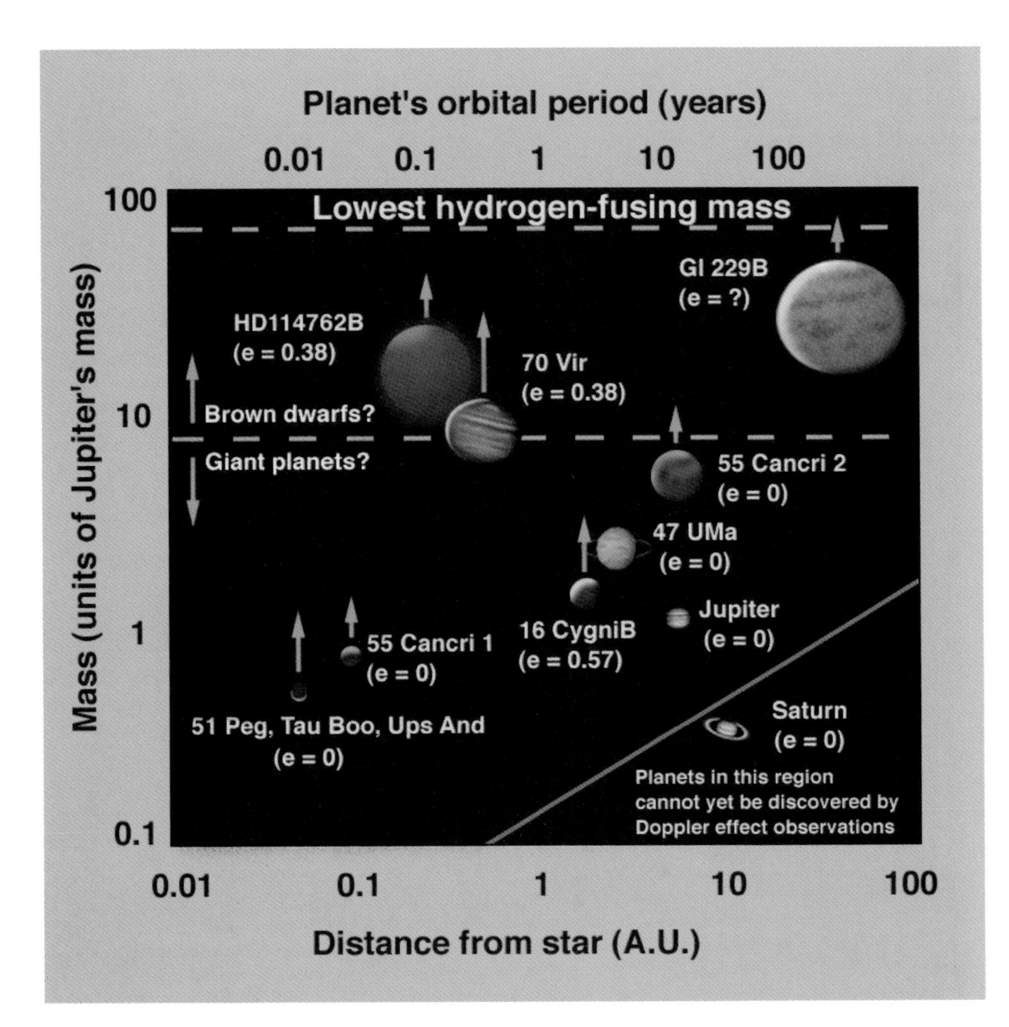

COLOR PLATE 22 This diagram, first created by the astronomer Alan Boss, plots the masses of different objects (vertical scale) versus their distances from the stars that they orbit (lower horizontal scale) and their orbital periods around a sunlike star (upper horizontal scale). The diagonal line shows the current limits on the detectability of objects like these by means of the Doppler effect: Only the objects above and to the left of the line produce detectable Doppler shifts.

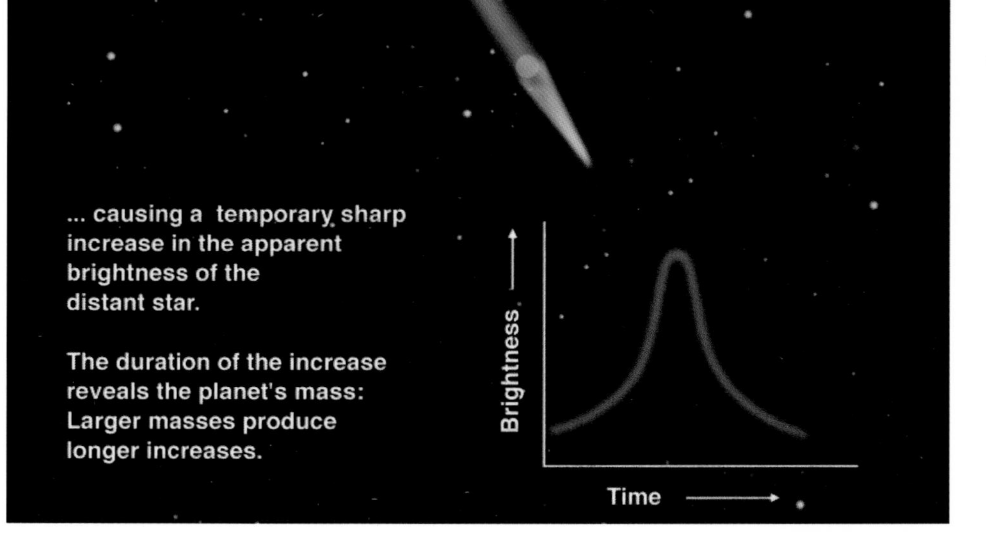

COLOR PLATE 23 An object's gravitational force bends light passing close by, creating a gravitational lens that focuses the light from a distant object and changes the object's apparent brightness. If astronomers observe only the change in brightness, they call this effect *gravitational microlensing*. They can use this effect to find objects that emit no light and are not otherwise detectable.

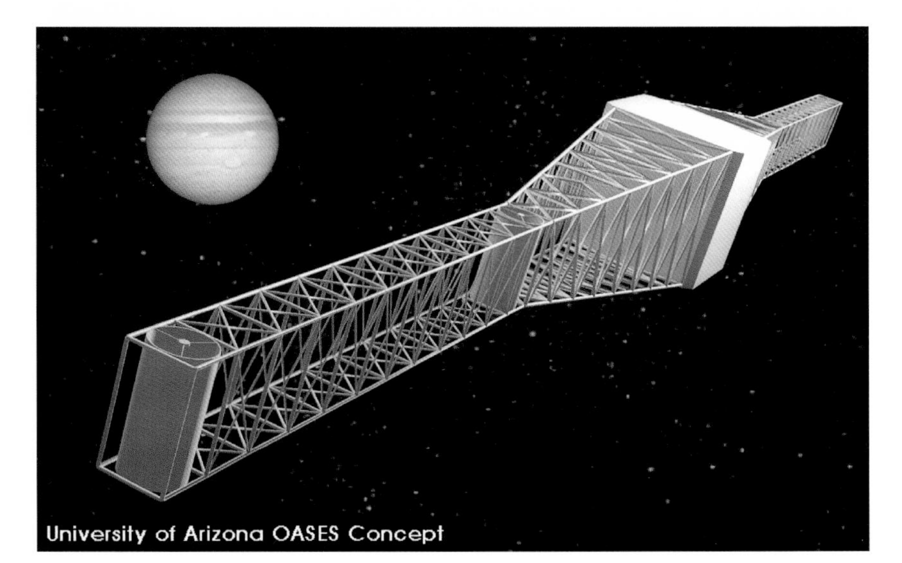

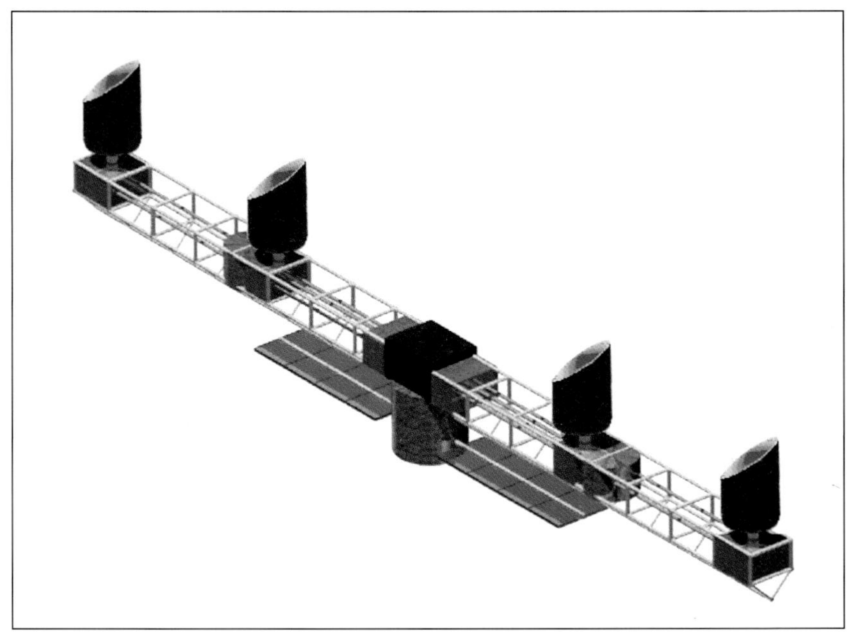

COLOR PLATE 24 These artist's conceptions show two possibilities for an interferometer system that could be sent out to 4-5 A.U. from the sun, to search for extrasolar planets without being overwhelmed by light reflected from particles of zodiacal dust. Further studies will be needed to determine how to design, construct, send, deploy, and maintain such an interferometer, which must keep an incredibly precise alignment among its component parts to use its interferometric capabilities. Such a system, many meters long, must assemble itself automatically after reaching its orbital destination.

the changes in brightness, called the "light curve," of any star whose brightness varied. With this equipment, each of the macho teams began to find events that spoke clearly of gravitational microlensing. Unfortunately, the microlensing effect produced in any single event depends on a combination of the mass, distance, and velocity across the line of sight of the lensing object. Observations cannot yield each of these parameters unambiguously, but with a significant number of events, astronomers can begin to get a statistical handle on the three parameters.

By now, two or three years after the macho-detection programs began, the astronomers have found several events, with characteristics typical of objects with masses between 1/10 and 1 solar mass, along the lines of sight to stars in the LMC. They have found about ten times more microlensing events when they look at stars in directions generally toward the center of the Milky Way. These results establish the practical usefulness of gravitational microlensing, at least for detecting objects with considerably more mass than even a giant planet's.

Amazingly enough, the distances to the least massive objects found by gravitational microlensing cannot be established within a billion light years! The objects that produce this microlensing lie along the line of sight to the quasar Q0957+561 A,B. *Quasars* are immensely luminous, pointlike objects, possibly the centers of young galaxies, that lie billions of light years away and therefore appear to us not as they are now but as they were billions of years ago. The "A,B" denotes a quasar for which gravitational lensing by an object between us and the quasar has produced a double image. From their decade-long observing program, astronomers at the Harvard Center for Astrophysics conclude that the quasar's modest changes in brightness imply gravitational microlensing by numerous objects with masses a few times that of Earth. These objects have masses far below the least massive objects found to date in the macho surveys, which, to be sure, have been designed to discover objects with much greater masses by gravitational microlensing. The microlensing technique gives no good indication of whether the low-mass objects found along the line of sight to the quasar lie close to the Milky Way, close to the quasar, or somewhere in between.

As the macho observations continue in the Southern Hemisphere, astronomers will make more, and more accurate, observations of distant stars. Eventually, astronomers should encounter cases where both a star and one of its planets pass almost directly between ourselves and another star. They should then observe a small blip superimposed on the shoulder of the light curve, which shows a long, slow rise and a symmetric long, slow fall, each lasting several weeks, as one star produces gravitational microlensing in the light from a more distant star. The blip would last for a few hours and should be easily visible as an exception to the smooth rise and fall. Since the probability of detecting a planet with this microlensing method varies in proportion to the square root of the planet's mass and does not much depend on the planet's distance from its star, gravitational microlensing should reveal the most massive planets most easily. However, as the quasar example shows, this method can reveal planets with masses comparable to Earth's.

The overall plan for using gravitational microlensing to find planets employs two medium-sized telescopes to monitor about 30 million stars each night (!), with the hope of discovering several hundred cases per year of microlensing early on the rising part of the stars' light curves, and to follow those stars especially closely to search for blips. If every star has a planetary system something like the sun's, every year this approach would yield a few dozen planets with Earth-like masses and a few hundred with masses similar to Jupiter's or Saturn's. Thus negative results over a few years would imply that planetary systems like our own must be extremely rare in the Milky Way.

Some astronomers find this technique for finding extrasolar planets the greatest thing since sliced bread. Others, however, interpose the serious objection that *detection of any particular planet occurs only once.* Gravitational microlensing ranks as an indirect technique for finding planets, since we see the planets' gravitational effects on the light from more distant stars. We know nothing about the planets' distances from us, save that they must be less than those to the stars we observe, and we may not even be able to see the stars around which the planets move in orbit. Certainly we can never—until our abilities increase markedly—hope to *see* the planets. For some, this makes the microlensing approach less appealing, but well-balanced reflection shows that we may well profit from discovering the existence of planets (or, for that matter, their nonexistence) in a statistical manner only. Gravitational microlensing has a fine future, one that would make Einstein proud, for in his day the strange predictions of his general-relativity theory seemed to have no practical impact.

One More Way to Find Planets: Contact with Extraterrestrial Civilizations

Contact with another civilization ranks high among the goals of those who seek to discover more about the cosmos, not only for its intrinsic interest but also for the useful and intriguing information that another civilization could furnish to us. As discussed in chapter 8, consideration of this subject comes loaded with emotion and prejudice, most of it deleterious in the struggle to gain a good perspective on the universe. However, no one doubts that if SETI, the *S*earch for *E*xtra*t*errestrial *I*ntelligence, should succeed, our perspectives would change drastically—at least until the next fall television season. At the very least, another civilization could tell us how distant their home (presumably another planet) lies from the solar system, and how they have assessed the fraction of stars that possess other planetary systems and possible abodes for life. Even the first contact will immediately set a rough calibration of the distribution of civilizations in the Milky Way, just as the first discovery of planets around other stars gave a rough idea of the fraction of stars with planets.

Of course, recognizing the benefits that would arise from another civilization's choosing to share its knowledge with us does not imply that we should abandon our own searches for planets around other stars. We are now in a position to use the knowledge we have gained to attack the issues that dominate our thinking about extrasolar worlds: How many are there? How many of those are habitable? How many actually possess life, and on which planets has life produced what we would call a civilization? By answering these questions, if only with algebraic symbols to stand for what we do not know, we can estimate the distance to the closest civilizations with which we might someday communicate. The next chapter of this book examines the numbers question, while the one after that deals with the potentially explosive issue of what form our interactions might take—worship, invasion, communication, or utter boredom?

CHAPTER

7

How Many Worlds in the Milky Way?

The discovery of planets orbiting sunlike stars in the sun's neighborhood provides a new perspective on the question of how many solar systems sprinkle the Milky Way galaxy. Can we determine whether most stars, or only a minority, have planets? Equally important from a human perspective, can we state whether most planetary systems contain at least one Earth-like planet, in addition to the giant planets revealed by the Doppler-shift observations of stars?

The Profits and Perils of Generalization

Astronomers draw their conclusions about the cosmos from a host of observations, dealing with objects whose classification ranges from "unique" to "so numerous that future generations will still be making catalogues." For this reason, a good working knowledge of statistics forms part of a trained astronomer's mental arsenal. As we all know, statistics ranks next to economics in producing stupefaction. Nevertheless, statistics demands respect in our quest to understand the prevalence of worlds. In particular, we must recognize that we *can* generalize from a relatively small sample, provided that we remain ever aware of two crucial sources of possible errors: The small sample *size* inevitably creates a potentially large error in our generalizations; and the *method* used to obtain the sample creates another, independent source of error. Let us never lose sight of these facts as we examine what we know about planets orbiting some of the Milky Way's 300 billion stars—an estimate obtained by sampling and other summation techniques, quite possibly in error by plus or minus 25 percent, or even more.

What Statistical Conclusions Can We Draw from the New Discoveries?

In chapters 1 and 6, we examined the Doppler-shift technique for finding new planets, as well as the analogous pulsar-timing techniques (which are much more accurate, and therefore capable of revealing much less massive objects, than current Doppler-shift measurements). We commented then that these methods favor the discovery of more massive and faster-moving objects, which produce greater *reflex effects* on the objects we observe. When we observe to a particular level of accuracy, we can expect to find only those objects above certain thresholds in mass and orbital velocity. Color Plate 22 shows these thresholds for astronomers' best efforts in Doppler-shift techniques, as of 1996. Note that both axes of the diagram are "logarithmic": The numbers along the axes increase by multiples, so that the diagram easily covers distances between 1/100 and 100 A.U. and masses between 1/10 and 100 Jupiter masses. Color Plate 22 demonstrates that Doppler-shift measurements can now reveal a planet with the mass of Jupiter, orbiting at 20 A.U. from its star—provided that astronomers amass observations covering a time period that includes most of an orbit. The detection of Jupiter-mass planets farther than 20 A.U. from their stars must await improved observational techniques. Since a planet at 20 A.U. from a sunlike star takes nearly a century for each orbit, we have time to wait for this improvement. For now, we must plan to use the "accessible" portion of the diagram, the part above and to the left of the shaded area, as a means of generalizing about the "inaccessible" portion. This will remain true even as improved techniques push the boundary down, so that the accessible portion above increases at the expense of the inaccessible portion below.

What do the results from 1995 and 1996 tell us about the abundance of planets in the Milky Way? *Of the nearby sunlike stars subjected to Doppler searches, about one in a hundred have shown a single planet in the accessible region of the diagram.* Because the Doppler-shift technique has reached a precision of 3 meters per second only during the past few years, astronomers have not yet had time to find planets orbiting their stars at more than a few A.U. from the sun. Our own planetary system now represents a system at the edge of detectability. If astronomers on a planet orbiting a star in the solar neighborhood happened to have duplicated exactly our developmental history (hardly likely, especially as to an exact congruence in time), then their measurements of Doppler shifts in the sun's spectrum for Doppler shifts would now have the capability, but would not yet have had the time, to find Jupiter. Another few years of observation would yield the discovery of the sun's largest planet, and a few more would reveal Saturn.

The first few discoveries of extrasolar planets allow the rather confident generalization that sunlike stars with planets detectable with our best current Doppler-shift techniques amount to no more than a few percent, and quite possibly only 1 percent or less, of all the stars classified as similar to the sun. We must not, however, think that this conclusion means that only a few percent of all stars have planets. Techniques other than Doppler shifts, such as the astrometric studies pursued by George Gatewood (see page 133) and the proposed space-borne transit surveys (see page 135), may soon reveal Jupiter-like planets orbiting at star-planet distances much greater than those of the planets that Doppler-shift observations can find. The indications of planets around Lalande 21185, one of the sun's closest neighbors, hint that a large proportion of all single stars may turn out to have planetary systems.

In chapter 4, we made another broad generalization: Single stars are far more likely to have planetary systems than double- or multiple-star systems. This conclusion remains largely drawn from theory and calculation, though astronomers will eventually test it observationally by devoting significant efforts to studying double stars. For now, the theoretical arguments seem sufficiently compelling that the astronomers who search for planets prefer to devote their scant resources of time and funding to the most promising candidates, lone stars like our sun. Ruling double- and multiple-star systems out of consideration excludes about half the stars in the Milky Way.

Some Stars Are Too Hot

Another wholesale exclusion occurs when astronomers concentrate their search on sunlike stars. Roughly speaking, sunlike stars constitute about 10 percent of all the stars in our galaxy. Another 5 percent of the stars in the Milky Way have significantly higher surface temperatures, greater luminosities, and shorter lifetimes than the sun. Any planets around these stars must be relatively young: No dinosaur gazing at the skies could have seen the stars that now mark the constellation Orion, or shine as the seven sisters of the Pleiades. Although they dominate the night skies, young, hot stars have less appeal to humanity as possible sites for intelligent life (see page 162). Furthermore, the stars' spectra contain fewer lines for careful measurement, and their rapid rotation broadens these lines, smearing them in wavelength (thanks to the same Doppler effect that also offers the chance to detect planets). This reduces the accuracy with which astronomers can measure the changes in the wavelengths of the spectral features. The difficulty of making accurate measurements, along with the relatively short lives of young, hot stars, leads astronomers to concentrate their searches on longer-lived, sunlike stars for which the most accurate measurements can be made.

Other Stars Are Too Cool

Fully 85 percent of the stellar population of the Milky Way consists of stars with significantly lower masses, cooler surface temperatures, and longer lifetimes than the sun's. Although none of these stars can be seen without a telescope, they dominate by numbers. Nevertheless, except for Barnard's Star (and now Lalande 21185), many experienced astronomers could not name a single one of these modest neighbors of the sun. The closest of these cool, dim stars bear names such as Wolf 359, Ross 154, and Groombridge 34, testimony to the patience of the eighteenth- and nineteenth-century astronomers who listed them in catalogues that carry the astronomers' names. If Lalande 21185 or any other of these stars has planets, those planets could bask in their stars' heat and light not for a mere 10 billion years, as the sun's planets will, but for 30, 50, 100 billion years or more. The down side lies in the cool stars' low luminosities, which are less than 1/200 of the sun's. These low luminosities imply that all but the closest planets will receive only modest amounts of light and heat.

Why have the astronomers who search for planets by the Doppler-shift method chosen to skip over this vast majority of stars, concentrating instead on sunlike stars? Most of the answer lies in the fact that the cooler stars tend to be noticeably dimmer than sunlike stars, despite the fact that they dominate the list of the stars closest to the sun. Their dimness, which arises from the cooler stars' lower luminosities, increases the length of time necessary to make accurate Doppler-shift measurements. In addition, a planet orbiting at a particular distance from a star will swing a lower-mass star at a smaller velocity around the center of mass. One might think that a low-mass star can be accelerated more easily—which is true!—but, on the other hand, a planet at a particular distance moves more slowly in orbit around a low-mass star, because the low-mass star likewise produces a smaller acceleration of the planet. The net effect favors the discovery of planets around higher- rather than lower-mass stars.

How Many Planets Are Implied by the Current Results?

For understandable reasons, astronomers have concentrated their Doppler-shift searches on the 10 percent of all stars that are roughly sunlike and have excluded more than half of those for multiplicity of the star system. The Doppler-shift result—that about one in a hundred sunlike stars has a massive planet orbiting relatively close to it—comes from studying a few dozen nearby stars, which statistically represent only about 5 percent of all the stars in our galaxy. A cautious extrapolation of the new results states that our galaxy appears to contain about 30 billion sunlike stars, of which perhaps 15 billion are single stars; of these 15 billion singletons, a hundred million or so possess a planet with a mass equal to at least half a Jupiter mass, orbiting at a distance less than about 2 A.U. In short, the new discoveries lead statistically to a rock-bottom conclusion that *many million sunlike stars in the Milky Way have planets*.

What can the other methods of searching for planets, described so thoroughly in the previous chapter, tell us about planetary numbers? Since the transit and gravitational-microlensing methods have barely begun, we may rank any attempts to use their (non)results as premature. The astrometry method, on the other hand, has been around for a long time and seems to have finally yielded positive results. If the announcement of planets around Lalande 21185 proves correct, and if we daringly extrapolate from a single planetary system found by this method, we can say that at least a sizable minority, and possibly the majority, of the Milky Way's cool, dim single stars have planets. On the other hand, Lalande 21185's planets may prove illusory, and the fraction of the cool, dim single stars with planets might be only a few percent, or even less.

Because the pulsar planets seem such an oddity, we shall exhibit a conservative strain by leaving them out of the calculation. In any case, as we shall see, pulsar planets hardly add to the potential sites for life, for they formed after the explosions that destroyed the pulsars' progenitors and have therefore existed for "only" a few tens or hundreds of millions of years. Even if the pulsar planets had miraculously preserved or acquired the chemical compounds and climatic conditions essential for life, this leaves little time for life to evolve into forms interesting to us. In the estimates we shall make about the number of civilizations that now exist in the Milky Way, we may regard pulsar planets as an extra, a type of planet that we can mentally keep in reserve without actually entering into the calculation.

If we estimate the total number of single stars in the Milky Way at 150 billion, we can now conclude that somewhere between 100 million and, say, 50 billion of those stars have planetary systems. The significance of this range of numbers lies in its lower, not its upper, bound. I have set the upper bound at 50 billion because I cannot believe that all single stars have planets, and we do have some negative evidence for the majority of single stars, the closest cool stars. But from 50 to 100 amounts to only a factor of two, whereas from 1, which before 1995 was an arguably reasonable total for the number of planetary systems in the Milky Way, up to 100 million represents quite a leap.

Does One Planet a "Planetary System" Make?

The preceding pages have built on a firm foundation of fewer than a dozen planets to make bold pronouncements about the millions of "planetary systems" in the Milky Way. We must now ask whether the discovery of a single planet, even one with more mass than Jupiter, implies that a planetary system exists around a star.

In 1995, before the announcement of the planet around 51

Pegasi, astronomers had an easy answer to this question: Yes. The standard model of planet formation discussed in chapter 4 envisages planets forming from smaller planetesimals, with each growing planet "owning" a particular ring of distances from the protostar through the influence of its gravitational forces. Planets that form in a specific orbit remain in that orbit, so that a planet's "ownership" of a particular range of distances lasts throughout both the epoch of formation and the billions of years afterward. To the extent that the standard model grew from a knowledge of our own solar system, it implied that the discovery of a giant planet at 5–30 A.U. from the star would signal the existence of smaller planets closer to the star, each of which had once gravitationally "owned" a set of distances from the star. An extrasolar observer who detected Jupiter, for example, would quite logically (according to the standard model) deduce that the sun must have a number of smaller planets in orbit—perhaps half a dozen, perhaps two dozen, perhaps even more. The one conclusion *not* logically accessible to that observer would be a lone planet that had acquired essentially *all* of the rotating disk available for planet formation.

Times change facts, and facts change ideas. Today we know at least five planetary systems (see Table 1.1) in which a planet with at least 0.6 Jupiter masses orbits its star at a distance less than 1/8th of the distance from the sun to Mercury. In chapter 4, we discussed the mechanisms that could produce this situation and concluded (rather, we allowed the experts to conclude) that these giant planets must have formed much farther (at least 4 or 5 A.U.) from their stars and must have changed their original orbits for much smaller ones. As the giant planets moved inward, their gravitational forces must have attracted all the matter moving in the orbits they slowly crossed. Hence an essential point about sunlike stars with giant planets close to them lies in the absence of Earth-like planets—indeed, of *any* planets between the star and distances greater than 4–5 A.U. These stars might possess other giant planets at still greater distances, similar to the distances from the sun to Saturn, Uranus, and Neptune (9.5, 19.2, and 30 A.U., respectively), and thus might have true planetary systems, even though none of their planets would resemble Earth, Venus, Mars, or Mercury. On the other hand, the giant planet that changed its orbit might have consumed nearly all of the protoplanetary material.

Resolving this issue will take significant further effort and serious expenditures of time, because the Doppler-shift techniques that revealed the close-in giant planets will require 25 to 100 years to reveal a Saturn-like planet orbiting its star at 10 to 30 A.U. We must therefore proceed with caution, remaining aware that in contrast to the implications of the standard model, the discovery of giant planets orbiting close to sunlike stars does *not* assure us that we have found a planetary system. This consideration plays an important role when we turn to the enjoyable speculation that accompanies an estimate of the number of worlds in the Milky Way: How many of them offer favorable sites for life?

Worlds and Habitable Worlds: The Drake Equation

We have spent sufficient time in gauging the number of planets in the Milky Way to reward ourselves with the consideration that drew us into this effort: How can we estimate the number of potential sites for life, and the number of advanced civilizations that now exist in our galaxy? An accurate estimate of the number of technologically advanced civilizations plays a crucial role in assessing the difficulty of establishing contact with other civilizations. In any statistically reasonable galaxy, the smaller the number of civilizations, the greater the distance to our closest neighbor civilization will be, with a consequent increase in both the difficulty of establishing contact and the time we must wait for messages or visits to pass between us.

In their estimates of the number of extraterrestrial civilizations, scientists use the *Drake Equation*, first developed by the astronomer Frank Drake, to summarize their knowledge and ignorance. To obtain its estimate, the Drake Equation multiplies different terms, each of which represents a condition believed to be necessary for a civilization to exist. We might compare this approach with an attempt to estimate the number of great restaurants in the United States, which takes the number of cities, multiplies this by the fraction of cities that have restaurants (nearly all, of course), and then by the average number of restaurants per city to find the total number of restaurants. This approach presupposes that great restaurants are found in cities, analogous to presupposing that planets are to be found around stars. If we multiply the number of restaurants by the fraction of all of them that have the decor that we seek, then multiply again by the fraction that have the food we enjoy, multiply once more by the fraction of restaurants that have the right prices, and then by the fraction where the personnel behave as they should, we obtain a product—much smaller than the total number of restaurants—equal to the number of restaurants in cities that have the right decor, the right food, the right prices, and the right personnel.

In a similar way, the Drake Equation starts the estimate of civilizations in the Milky Way with an estimate of the number of stars in our galaxy, $\mathbf{N_s}$. If we expect to find life only on planets that orbit stars that last for at least a billion years, then we must multiply this number by the fraction of all stars that are roughly sunlike in their lifetimes ($\mathbf{f_s}$) and continue by multiplying this product by the average number of planets per sunlike star ($\mathbf{N_p}$). This gives us the total number of planets orbiting sunlike stars. Of all these planets, some fraction (call it $\mathbf{f_e}$) provide suitable places for life to develop, mainly, we may conclude, because of the temperature requirement that we discussed on page 32. The product of the first four factors ($\mathbf{N_s} \times \mathbf{f_s} \times \mathbf{N_p} \times \mathbf{f_e}$) provides the number of planets suitable for life in the Milky Way. If we multiply this product by the fraction of planets suitable for life on which life actually does develop, which we may denote as $\mathbf{f_l}$, and then multiply by the fraction of planets with life on which civilizations arise ($\mathbf{f_c}$), we find the total number of planets in the Milky Way that have a civilization at some time in their history.

However, we are interested mainly in the number of civilizations in the Milky Way *now*. In this case, provided that our epoch can be considered as roughly representative—that is, much like the Milky Way a few billion years ago or a few billion years in the future—we must multiply the product we have created by a final factor: the ratio of a civilization's average lifetime to the lifetime of the Milky Way. This produces the following result for N, the number of civilizations in the Milky Way today:

$$\mathbf{N} = \mathbf{N_s} \times \mathbf{f_s} \times \mathbf{N_p} \times \mathbf{f_l} \times \mathbf{f_c} \times \left\{ \frac{\text{lifetime of average civilization with communications ability and desire}}{\text{lifetime of Milky Way galaxy}} \right\}$$

When Frank Drake first wrote this equation, nearly four decades ago, he was a young astronomer at the National Radio Astronomy Observatory in West Virginia. At that time, astronomers had a good knowledge of $\mathbf{N_s}$, the number of stars in the Milky Way, and of $\mathbf{f_s}$, the fraction of stars that last as long as the sun, but they had essentially no knowledge of any of the terms farther to the right in the equation. The discovery of planets in the Milky Way can be summarized, once we have some additional planets and planetary systems, as "one more term in the Drake Equation," a further step toward an accurate estimate of the number of civilizations that now exist in our galaxy. Just four more steps will then give us the answer we seek. If each step requires 40 years, our grandchildren's grandchildren may yet know the first five terms that enter the calculation of **N**. The final two terms, which involve the fraction of planets with life where civilization develops and the average lifetime of a civilization, cannot be determined accurately until we establish contact with another civilization. This contact could, of course, provide an alternative to the dogged scientific research otherwise necessary to determine the earlier terms in the equation.

Suppose that, despite our ignorance, we estimate all the factors in the Drake Equation save the last one. The number of stars in the Milky Way is close to 300 billion, and the fraction of those stars that last at least a few billion years is close to 0.95. The number of planets per star might be as low as 0.01. This would be true if most planetary systems resemble those revealed so far by Doppler-shift techniques, if other techniques fail to reveal significant numbers of farther-out planets around sunlike stars, and if the majority of stars—the red and orange stars, fainter than the sun and its sunlike cousins—have the same planetary characteristics that sunlike stars do. On the other hand, the number of planets per star may yet prove to lie between 1 and 10, if most single stars have planets (most of them presumably too small, and too far from their stars, to be discovered around nearby sunlike stars by the Doppler-shift method) and if the standard model has overall validity for most planetary systems. Let us, in order to proceed with our speculative process, set the number of planets per star, N_p, equal to 0.1, while remaining sensitive to the algebraic truth that each factor in the Drake Equation multiplies all the others, so that a change in one factor by a factor of ten will change the result by the same factor.

What fraction of all planets occupy a "habitable zone" or are otherwise suitable for life? The standard model and the example of the solar system would suggest numbers between about one-ninth (if we judge only the Earth to be habitable) and two- or three-ninths (if we add Venus and Mars as marginal candidates, or perhaps allow large satellites to rank as planets and find one or two of them habitable). On the other hand, if the new planets represent the norm among planetary systems, then the fraction of planets judged habitable may fall precipitously: Neither the close-in giant planets nor any other giant planets at distances of 10 A.U. or greater seem likely to offer habitable conditions. On the third hand, let us not forget that two of the new planets, the ones around 47 Ursae Majoris and 16 Cygni B, orbit more or less within the habitable zone. A conservative choice for the fraction of planets with conditions suitable for life, f_e, might be about 0.1, with alternative justifications. Most planetary systems may turn out to resemble the solar system more closely than the 51 Pegasi system, in which case f_e should be at least 0.1; on the other hand, if the first crop of new planets to be discovered provides a good sample of all systems (though we must never forget the bias introduced by the Doppler-shift method!), then at least one of the first ten has a planet in the habitable zone around its star.

On what fraction of planets suitable for life does actually develop? Data to date suggest an answer close to unity; conservatism prompts an answer closer to 0.5; pessimism leads us below 0.1; and a taste for regarding life on Earth as a miracle speaks for 0.0001 or less. Let us set the fraction of suitable planets on which life does develop, f_l, at 0.2 and move on to f_c, the fraction of planets with life on which "civilization" evolves. We may quarrel about which societies deserve this title, but from an astronomical view, a "civilization" means, without value judgment, *a society capable of sending messages across interstellar distances*. By this definition, Earth has had a civilization for about a century. Notice that assigning a small value to f_c implies that our Earth somehow stands out among planets with life. In opposition, consider the "argument from mediocrity," that is, from the concept that our own history provides a roughly representative example of what happens on a planet with life. This suggests that where life exists, evolutionary processes will lead to what we call intelligence, and then to civilization, so that f_c should rise to close to one. Suppose we try $f_c = 0.5$ and turn to the last term, the ratio of an average civilization's lifetime to the (roughly) 10-billion-year age of the Milky Way.

The hundred years of the twentieth century might turn out to include half our planet's lifetime with interstellar-communication ability. More optimistically, they could represent no more than, say, one part in a million of what turns out to be the Earth's total lifetime (100 million years!) with a civilization, as we have defined it, in existence. Can we use our society to estimate not only the future prospects for civilization on Earth but also the past, present, and future of other civilizations in the Milky Way? For once, let us be utterly, if temporarily, conservative, and retain the symbol **L** to represent the average time interval during which a planet possesses a civilization that can send or receive interstellar messages.

What, then, does the Drake Equation tell us? Using the numbers we (in this case, the author) have found reasonable, we now have

$$\mathbf{N} = \mathbf{N_s} \times \mathbf{f_s} \times \mathbf{N_p} \times \mathbf{f_l} \times \mathbf{f_c} \times \left\{ \frac{\text{lifetime of average civilization with communications ability and desire}}{\text{lifetime of Milky Way galaxy}} \right\}$$

$$= (300 \text{ billion}) \times (0.95) \times (0.1) \times (0.2) \times (0.5) \times (\mathbf{L}/10 \text{ billion years}).$$

If we measure **L** in years and perform the multiplication, this gives us **N** = 0.0283 **L**, which we may round off to **N** = **L**/35. This means that if—just a modest "if" to summarize the realms of speculation in which we indulge with the Drake Equation—our numbers are correct, then a short average lifetime for a civilization (100–200 years) implies that the Milky Way at any given time contains only a few civilizations, whereas a long average lifetime (say, 1 million years) for a civilization that has developed interstellar-communications ability indicates that **N** measures in the tens of thousands. If we regard the apparent fragility of our present civilization as a temporary phase and recall that dinosaurs and cockroaches each lasted for more than 100 million years, we may even contemplate **N** reaching into the millions.

Suppose that our estimates are at least vaguely accurate, so that **N** does approximate **L**/35. What does this imply for interstellar communication? The result means that if **L** turns out to be a few thousand years, we would expect to find only a few dozen civilizations in the Milky Way at any representative time, such as the present. In that case, the average distance between neighboring civilizations will equal thousands of light years. In contrast, if **L** turns out to be hundreds of million years, then at any time the Milky Way should contain millions of civilizations, and the distance between neighboring ones should be only a few dozen light years. In the intermediate case of **L** equal to a few hundred thousand years, **N** would equal about ten thousand, and the average distance between neighbors would be a few hundred light years.

Note that among all the other disclaimers we have managed to introduce, **L** denotes the average lifetime of a civilization with both the ability and the *desire* to communicate. Both of these qualities appear to be essential requirements if we hope to establish contact. Highly advanced civilizations may have no interest in making contact with infant societies similar to ours. We may draw the general conclusion that we are most likely to discover first the civilizations more like our own.

So: Where Is Everybody?

We can slice and dice the numbers in the Drake Equation however we like without satisfying the skeptics who ask, as the great Italian physicist Enrico Fermi did when the issue arose, Where is everybody? If all the fractions in the Drake Equation exceed 1/100, and if **L** exceeds a few thousand years, the Milky Way should have several civilizations at least. Why haven't we heard from any others? (We deal with the explanation that we *have* heard from them—and in fact have been repeatedly visited by them—in the next chapter.)

Fermi seems to have been just a trifle quick in his criticism. If the Milky Way contains a thousand civilizations, the closest of them should lie about a thousand light years away, more than 10 million times more distant than the planet Jupiter. A society might develop through many centuries before finding the desire, and the means, to embark on journeys this long, or even to send deliberately beamed messages into space that signal their existence. Of course, if other civilizations use radio waves as we do, for radio and television broadcasts and for radar location, then they inevitably radiate into space a totality of electromagnetic signals that could be detected as non-natural in origin. This thought underlies current efforts in SETI (the Search for Extraterrestrial Intelligence), which survey the sky partly in order to find “leaked” broadcasts. However, because SETI searches have barely begun, the lack of positive results has few implications for the prevalence of extraterrestrial civilizations.

Those who believe in extraterrestrial visitors to Earth solve this problem by letting the other guys do the work. If instead we remain skeptical of claims of extraterrestrial abductions, alien-made crop circles and animal mutilations, and other supposed manifestations of the mindset of otherworldly visitors to Earth, we can recognize that finding extraterrestrial life will require more serious efforts than waiting for attack or amusement from outer space. A deliberate, highly sophisticated search for signals from extraterrestrial civilizations would require an expenditure amounting to about one first-class stamp per household per year—about the cost of a feature film. (But isn’t that the argument made by logrollers of all stripes? Nevertheless it remains a sobering comparison to those who feel that this effort deserves funding.) Since the United States Congress has already contemplated a similar program, funded it during one year (1992–93), and then canceled all governmental support for the search for other civilizations, we cannot expect that the postage-stamp comparison will soon yield fruit. For now, we can examine the prospects for finding life elsewhere, not so much by deliberate effort but as part of the search for planets around other stars.

CHAPTER

8

Can We Find Life on Extrasolar Planets?

Life! We all have it, all want it, all have opinions on its meaning. And almost all of us have strong feelings about the abundance of life in the cosmos. Even more surprising, most of us have firm convictions about the degree of interest demonstrated by other forms of life toward our own life on Earth. For anyone who seeks seriously to discuss the probability of life on other planets, the first order of business—even before examining the origins of life and possible habitats for its further development—lies in dealing with the intuitive human belief that our planet plays a special role in the cosmos.

The belief that Earth forms the center of the universe springs from deep roots in human consciousness. Based on direct experience, this belief has played a useful role in human survival and cannot be changed by mere thought. Deep inside each of us lies a Ptolemaic conception of the heavens, in which a motionless Earth lies at the center of the system of sun, moon, planets, and stars that surrounds us. This conception can and does survive all the advanced learning and philosophy we may add to our store of knowledge; even the most highly trained scientist "knows," deep in his reptile brain, that the Earth is by far the biggest and most important object in the universe.

Intuition, Friend and Foe to Understanding

On the other hand, learning also plays a key role in human development. The acceptance of the Copernican doctrine, which dethroned the Earth from its central position in creation, represented a crucial step in understanding the cosmos—a step that most of the human population has yet to take. By now, more than half of the adults in the relatively well-educated United States and Europe will state, if asked, that the Earth orbits the sun rather than the reverse. We may regard this figure either as a testament to the fact that knowledge does diffuse successfully or as a warning that many of us will prefer to follow their intuition when considering the basic facts of modern science.

When we turn to the question of life beyond Earth, human intuition seizes control of our thoughts and can only with careful attention be held in check. This is not to knock intuition, which has helped us to reach our present stage of understanding and plays a crucial role in daily survival. But consider the first question you would ask if you heard that astronomers had found life on another planet: "What does it look like?" This question springs from our traditional ways of examining and understanding our surroundings; although an important one from a scientific perspective, the crucial questions to which scientists would seek answers have a different focus. Intuition requires that before we can comprehend an event, we must form a mental picture of it, and the more closely that picture resembles what we know, the better we expect our understanding to be. Quite naturally and quite rightly from a survival viewpoint, intuition therefore builds on the familiar while mistrusting and downgrading the unfamiliar. When intuition cannot place an event into a familiar category, it usually judges it to be either nonexistent, insignificant, or miraculous. Though miracles receive great psychic respect, they owe their emotional power to their place outside the frame of the familiar, that is, outside intuitive understanding. This is simply the flip side of insignificance: In both cases the event's distinguishing characteristics lies in its failure to fit into what we already know.

Consider how we apply our intuition to the question of extraterrestrial life. Most of us have absorbed the notion that stars sprinkle the heavens as thickly as grains of sand on the beach—a fact that does not contradict our intuition unless we pause to note that each of these stars generally resembles our own sun and far exceeds Earth in size. If many of these stars have planets, and if many of these planets are something like Earth, then we might reasonably expect that the cosmos contains myriad planets with life. If we paused to consider how insignificant this makes our own civilization, we might be appalled; in 1600 Giordano Bruno was burned at the stake in Rome for insisting that God must love these other beings as much as ourselves. But instead of imagining the complex details of these other, hypothetical civilizations, our intuitive reaction typically asks, What do these civilizations mean *to us*? and answers, Why, nothing—until they come to visit.

In other words, our intuition, always looking out for our best interests, has combined a visually evident fact, that the night sky is spangled with lights, with a viscerally obvious one, that we humans are so special that Earth must be the center of the universe. At a more cerebral level, the interplay runs as follows: We have learned (so say our advanced brains) that our Earth is nothing special, that other stars by the billions probably have planets too and are therefore likely to have life around them. And then, skipping over the fact that makes this possible—the immense size of the universe, which involves enormous distances between neighboring stars, a fact completely hidden from direct perception—our reptile brains leap to the obvious conclusion that this multitude of life forms must take a tremendous interest in us. They must want to check us out, for who would not?

Let me be clear: I too value intuition, use it constantly, and admire its abilities. But when we seek actual answers to our questions about life in the universe, we must start by recognizing the problems that arise when we move to situations beyond those that intuition can handle. Intuition does not disappear (it could no more do so than we could cease to breathe); instead, we then let it slide a bit farther underground in our consciousness.

UFOs and the Human Mindset

Consider, for instance, the numerous reports of extraterrestrial visitors to Earth. During bygone eras (I refer here to my youthful years, roughly the period from the Second World War through Dwight Eisenhower's presidency), these reports were characterized as UFOs, unidentified flying objects. Until the aerial bombings of Guernica in 1936 and Rotterdam in 1940, attack from above had seemed a fantasy, confined to books such as H. G. Wells's *War of the Worlds*; a few years later, from 1945 onward, the greatest threat to humanity's survival consisted of aircraft-borne (later rocket-borne) nuclear weapons. Unsurprisingly, when a private pilot named Kenneth Arnold reported saucer-shaped objects flying at immense speed near Mount Rainier in 1947, the Western world was primed to see danger lurking in what might be visitors from other worlds. UFOs had a direct emotional connection to hostile aircraft and were seen in that light: Our intuition had placed UFO reports in the most relevant accessible category—interplanetary intruders whose hostility resonated with that received from the menacing Soviet empire.

Eventually, we learned to live with the Cold War, and as the Cold War came to an end, UFO reports began to lose their emotional appeal. No longer did we hear of bright lights moving at incredible speeds (after all, a newspaper can print just so many of these reports, let alone their prosaic explanations described below); instead, a new connection with outer space emerged: the personal visit. Although the first report of such an encounter appeared in 1960, the 1980s and 1990s have seen a large increase in the number of persons reporting face-to-face meetings with alien beings, in many cases involving abduction to their spacecraft for medical examination or similar purposes. Assembling the "anecdata" of such encounters lies beyond my province as a science writer, though I often pause to envy the popular success of those who engage in his endeavor, but even a cursory reading makes it clear (to me!) that the phenomenon is definitely real: Most of those who report abductions are neither lying nor befuddled. Only the *objective* reality of their experience remains in doubt.

From a scientific viewpoint, this doubt nearly vanishes. Extraordinary claims demand extraordinary proof, and the reports of abductions by alien beings offer no proof beyond human recollection. For historical and practical reasons, our legal system often finds such evidence reliable, though hardly the most convincing type: One videotape of, for example, police officers beating a suspect carries far more weight than a dozen eyewitnesses. Notice that here our intuition leads us in the right direction. We tend to trust what we ourselves *see* more than anything another person tells us, partly because we have learned that events as related by another must pass through the filter of that person's consciousness and recollection. Thus we trust a videotape—but we fail to remember two crucial facts. First, filming must, by its nature, be selective in its timing, its subject matter, its focus, its depth of field, and a host of other variables. Second, even more important in the modern world, fakery in filming can produce images just as "real" and even more emotional moving than anything we encounter in our daily lives. Our conscious minds know that advertisers have built on these facts for many years, but we nevertheless remain sensitive to "powerful" images; indeed, for many of us, only the most carefully crafted filmic moments have much ability to move us.

However, the world in which we live continuously presents us with disjointed, unedited stimuli that we must constantly assimilate and attempt to understand. As we do so, our brains reconstruct reality in a way that harmonizes with past experience. Every police officer and criminal attorney knows the unreliability of eyewitness testimony, which can be far more pronounced when a witness tells "the truth" than if he or she attempts to lie. Even the best-trained witnesses, subject to emotional stress and an unusual situation, prove unable to recall or relate it correctly. This should come as no surprise once we recognize that our intuition functions by attempting to fit experience into familiar categories.

UFOs provide an excellent example of this category fitting. The vast majority of objects reported as UFOs turn out to be celestial objects, with Venus leading the pack. At certain points in its orbit, when Venus comes close to its minimum distance from Earth, the planet glows far more brightly than usual and reaches a maximum angular distance from the sun on the sky. Venus therefore remains longer in the skies of twilight, either before dawn or after dusk, than usual. Many an observer has noted Venus as an interloper, not present in memory, and has leapt to the conclusion that this alien object in the sky seems to menace the Earth. Consulting a star chart will verify that this is no familiar celestial object, for star charts do not show the planets—their continuous motions with respect to the background of stars would render such an effort useless. Finally, if a person happens to leap into an automobile and drive off to report the intruder, Venus will "follow" the car as surely as any alien intelligence could. The lengthy list of those whom Venus has fooled includes Jimmy Carter, who filed a UFO report before his campaign for the presidency in 1976, recording a mysterious object at Venus's location on the sky. Like Venus, Mars and Jupiter have figured in UFO reports, thanks to their changing apparent brightnesses. Mars's apparent brightness varies through an even wider range than Venus's as the planets orbit the sun, so it can become astonishingly bright (though not so bright as Venus) over a few weeks' time.

Not all the celestial objects reported as UFOs turn out to be planets. In fact, the best cases for demonstrating that even highly trained observers can and do make enormous mistakes when unexpected events occur lie in the UFO reports generated by meteors and other debris high in our atmosphere. On July 24, 1948, the crew flying an Eastern Airlines DC-3 on a clear night near Montgomery, Alabama, suddenly saw what seemed to be a giant aircraft approaching from the east. The aircraft passed by them at an estimated distance of 700 feet and a speed of 500 to 700 miles per hour; its size was estimated as about 100 feet long and 25 to 30 feet across (Figure 3). Both the captain and the copilot of the DC-3 described the object as having two rows of windows that appeared lit from within, and the captain stated that "You could see right through the windows and out the other side." The object was in view for about 10 seconds and then "pulled up with a tremendous burst of flame out of its rear and zoomed up into the clouds." The Air Force investigated this event carefully; one of their consultants, the astronomer J. Allen Hynek (later famous as a proponent of the belief that some UFOs represent alien spacecraft) wondered whether "the immediate trail of a bright meteor could produce the subjective impression of a ship with lighted windows."

Although the captain and copilot stated that they could not have mistaken a meteor some 30 to 50 miles high for an aircraft passing at 700 feet, exactly this seems the likeliest explanation of what happened that night over Alabama. Before passing to the example that makes this explanation most plausible, let me pause to underscore two facts. First, a scientific rather than an intuitive approach lays down the rule that the stranger the alleged explanation of an event, the stronger the proof must be to have that explanation accepted. Intuition does not allow the complexity embodied in this rule. Second, an honest photograph (but we must always bear in mind that establishing its honesty may be difficult) can outweigh all the honest eyewitnesses one can assemble. When Hynek taught introductory astronomy at Harvard during the late 1950s, he enjoyed describing his experience as a consultant to the Air Force and would whip a miniature camera from his breast pocket, proclaiming that without hard evidence as opposed to eyewitness testimony, one could never hope to obtain an accurate understanding of what had actually occurred. "If I ever see a UFO," he vowed, "I'm going to photograph it." With that statement, Hynek showed an excellent understanding of how science works, though he never did obtain an actual photograph of a UFO.

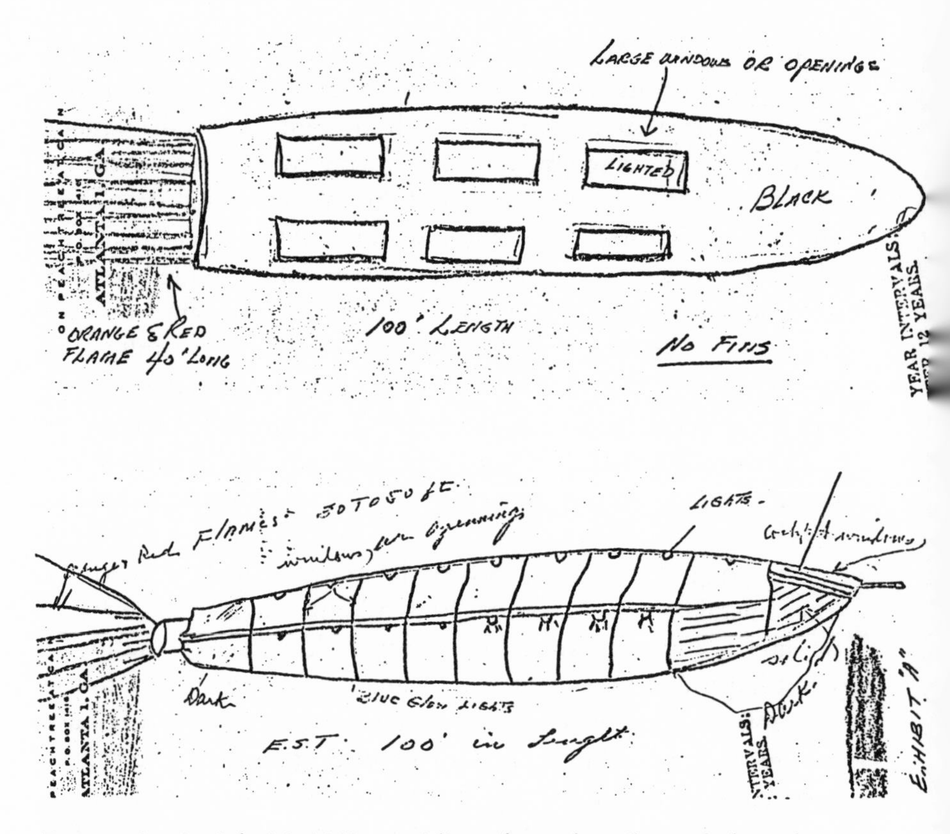

FIGURE 3 On July 24, 1948, an airline pilot and copilot each drew the UFO they had recently observed. Since they had previously discussed the appearance of the UFO, the divergence between their drawings seems remarkable.

Nearly 20 years after the UFO seen over Montgomery, a similar event occurred not far away. On March 3, 1968, three reliable witnesses in Tennessee saw a bright object that spouted orange-colored flame from its rear and passed silently overhead, no more than 1,000 feet above the ground, "like a fat cigar," as one witness put it. The UFO had at least 10 large, square windows apparently lit from inside (Figure 4). At the same time, 200 miles to the north, six observers in Indiana saw a cigar-shaped

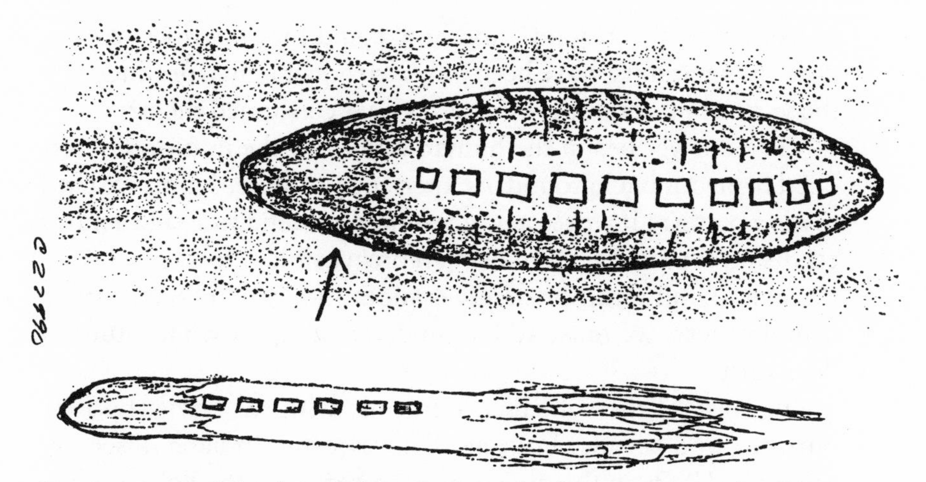

FIGURE 4 On March 3, 1968, a Soviet rocket booster grew so hot from its passage through the atmosphere that it glowed brightly and was drawn by observers in Tennessee (top) and Indiana (bottom).

object with many windows, described as 150 to 200 feet long and passing silently at treetop level. Observers in Ohio saw three similar objects, also silent.

In this case, we know the proximate cause of the UFO reports. On the previous night, the Soviet Union had launched a rocket, and one of its boosters fell back to Earth at the time of the UFO sightings. Just like the stony-iron objects that produce meteors as they encounter our atmosphere, the rocket booster grew steadily hotter as it passed through progressively denser layers of air, until it began to glow, split into fragments, and was eventually consumed by the heat induced by its passage through the atmosphere. Like a meteor, the rocket booster glowed brightly at altitudes of thirty to fifty miles, so it could be seen simultaneously by observers two hundred miles apart, for whom the various fragments apparently produced the effect of windows. But how could these independent observers, several of them sober citizens with no apparent inclination to visualize strange aircraft, mistake an object's altitude by a factor of a thousand, placing it only hundreds of feet above the ground? And how could they see rows of windows in what must have been only a string of lights, produced by the various pieces of the booster as heat consumed them?

The answer lies in the human mind—in the intuition developed from our background of experience. If we see lights, we tend to see a pattern; if we see a bright object, we tend to assign a modest distance to it. Far from being reprehensible (I feel I must state this again), our intuition deserves admiration and respect, but for completeness, we must add an understanding of what intuition does for us.

Because of the coincidence in time with the Soviet booster burn-up, and because both the Tennessee and Indiana observers saw their UFOs in the direction expected from the known location of the booster, the UFO reports from March 3, 1968 can be conclusively explained as sightings of a human-made meteor. The UFO from July 24, 1948 cannot be so certainly explained, but a comparison with the reports from the 1968 event, plus the fact that July 24 marks the peak of the annual Delta Aquarid meteor shower, makes a similar explanation quite attractive—from a scientific rather than an intuitive viewpoint.

Of course, bright planets plus meteors cannot explain all UFO reports satisfactorily. In our attempts to assess what UFO reports have to tell us, however, they throw an excellent light on the role of intuition and the untrustworthiness of eyewitness accounts. An additional point about meteors deserves notice: A particularly bright meteor, called a fireball or a bolide, caused by the passage of a relatively large object through Earth's upper atmosphere at speeds of 10 to 30 miles per second, will often leave a trail across the sky. When a fireball appears, eyewitnesses often hear a loud sound and see the object fall "just over the next hill." Again we see our intuition's attempt to fit the event to the expected pattern: Sudden bright lights usually have accompanying loud sounds, and objects that sweep rapidly across the sky "must" be close to us. We know that fireballs typically do not produce sounds, because sound-recording devices are often running at the times when they are photographed. Intriguingly, post-fireball interviews show that the longer the time interval between the event and the report, the larger the probability becomes that the fireball will be remembered as having been accompanied by a loud bang.

The well-balanced reader may conclude that most UFO reports based on eyewitnesses have relatively prosaic explanations, but a small residue remains that can*not* be so explained. Indeed this is so. The question then becomes one of choosing among alternative possibilities: the eyewitness(es) have significantly misremembered what they saw and heard; the eyewitness(es) are, in some unhappy cases, lying (this becomes more likely when, for example, a witness was abducted by aliens while late for work); the eyewitness(es) have in fact encountered an extraterrestrial spacecraft. In assessing these possibilities, we should always recognize the role that our intuition plays in processing what we observe. As with the recent discovery of possible ancient life on Mars, we would do well to maintain a conservative approach (for instance, that life has *not* been found on Mars and that extraterrestrial visitors have *not* reached Earth) until sufficient evidence to reject these views has emerged.

Alien Abductions?

What of the growing number of reports of abduction by alien visitors to Earth? Even if we assign mendacity to some of those who report abductions (I am reluctant to allow people to claim the title "abductee" by assertion, and "alleged abductee" seems a bit labored), no real doubt exists that hundreds or thousands of these reports come from people who are not lying, that is, not consciously telling a falsehood. John Mack, a Harvard University psychiatrist, has interviewed many people with abduction reports and stakes his reputation on his ability to recognize that they are telling the truth as they recall it. Unfortunately, Mack misses the point about intuition and concludes that he has heard objective reality. In contrast, I assert that a scientific approach—that is, one that relies on organized skepticism—requires far more proof than eyewitness accounts before one can reasonably conclude that these people have been taken aboard alien spacecraft. This skepticism serves the advancement of knowledge, which can proceed only by testing both evidence and the hypotheses that seek to explain the evidence.

But what to say about the general similarity that runs through these reports? Most of those who say they recall being abducted by aliens state that the aliens resemble humans in their overall structure, with small, slim bodies, large heads, and extremely large eyes. How can people who surely do not know one another make such similar statements—unless they have witnessed similar events?

To a scientist, logic suggests that the similarity of the descriptions of extraterrestrial beings argues strongly for a human rather than an extraterrestrial cause. To assume that such beings must resemble humans, even in the most approximate manner, is to fall into a Ptolemaic trap whose snares can be illuminated by a moment's consideration of the immense variety of life forms on our single planet Earth. Octopi, crustaceans, earthworms, and crocodiles do not much resemble one another. The similarity of appearance in the reported aliens seems far likelier to have arisen from a shared concept of what aliens ought to look like, a concept supported by a thousand movies and television programs, in which (at least until the most recent times in computer-generated graphics) paying actors to dress up as aliens costs far less than letting an art director's fancy roam freely. This explanation gains force from the undisputed fact that far more reports of abductions by aliens arise in the United States than in all other countries combined.

Perhaps the highest drama in the consideration of visitors to Earth lies in the assertion that the answer has long been known, but concealed from the public. Who has not heard that the United States government recovered bodies from a crashed alien spacecraft near Roswell, New Mexico, as early as 1947, has kept them on ice in a mysterious hangar ever since, has known of repeated visits to Earth by extraterrestrials, and has chosen to conceal this fact from us because our leaders know that the news would so perturb society that . . . we would no longer trust them to the point of reelecting them? This explanation assigns such stupidity to our leaders that one wonders how they achieved their electoral victories; a more appropriate reaction to an alien landing would be a call to spend enormous amounts on countermeasures that would reinforce our leaders' power and prestige.

The short answer to the assertion of hidden bodies consists of the fact that our government has repeatedly denied it. However, the time has long passed when this could be taken as final; at present, many citizens take this denial as the most eloquent form of confirmation. A longer answer refers to the fact that no conspiracy seems to have remained secret for long. From Catiline to Benedict Arnold, from cornering the gold market to Watergate, the various pressures to reveal secrets have soon proven victorious over the need to conceal them, and all the more rapidly when larger numbers of individuals must keep a secret. (The counter-argument is that the public has yet to know about the conspiracies not yet revealed.) In lectures, I like to point out that in view of the books and articles by which I have informed the public about the possibilities of extraterrestrial life, any conspiracy to conceal actual knowledge of alien visitors to Earth must inevitably include me, so any explanation I provide must be suspect. The fact that this usually provokes laughter among my listeners reminds us that conspiracies can succeed only by limiting their participants to a tiny number of individuals who know how to keep a secret, but the aliens-on-ice conspiracy would require large numbers of conspirators: those who found the bodies, those who transferred them, those who store them, those who told our leaders, the leaders who replaced those leaders, and so forth.

As long as I am launched on conspiracy theories, let me introduce my favorite, just as impossible of complete disproof as the dead aliens in the hangar. Could everyone on Earth except yourself be an extraterrestrial, pretending to act "normal" in an attempt to find out what a human is really like? What evidence exists in favor of this theory? What against it? What evidence could you hope to gain that would settle the question once and for all?

These are good questions to ask of any theory. If you find that you have a small bend toward paranoia, it might be worth asking, Who profits from promoting a particular answer? Could movie producers find it more profitable to make films about invasions by hostile aliens, rather than about aliens who can barely care what happens on Earth? Could citizens and even authors who promote abduction stories be influenced, consciously or unconsciously, by the fame and fortune they may receive? Could even apparently skeptical writers actually be preparing for a public conversion, in which they accept UFOs as extraterrestrial spacecraft?

The final insight I offer on the unreliability of eyewitness testimony comes from one of my favorite movies, the Marx Brothers' classic *Duck Soup*, in which the formidable lady played by Margaret Dumont spots Chico Marx in her bedroom and identifies him as the man she has seen leaving a moment before. Chico denies it, provoking Ms. Dumont to proclaim, "But I *saw* you!" Brilliantly Chico counters, "Who are you going to believe—me, or your own eyes?" And in fact, for reasons that need not detain us here, at this point in the movie Groucho, Harpo, and Chico have all donned identical nightcaps and nightgowns: The man Ms. Dumont saw leaving was Harpo. Probably the Marx Brothers neither knew nor cared that Chico was actually telling the truth in this scene, but his throwaway line eloquently summarizes the difficulty of knowing what to believe from eyewitness testimony.

Tradition demands that discussions such as this one lead to the famous statement by J. B. S. Haldane, the iconoclastic biologist, that "the universe is not only queerer than we suppose; it is queerer than we *can* suppose." (By the way, how did he know that?) Haldane did not, however, introduce his conclusion as an excuse for turning off one's mental processes when confronted with unusual, not to say bizarre, situations. Instead, one can make a good argument that the stranger the situation, the more one should remember to employ all the rational, skeptical powers one has, to avoid being snookered by emotion or snared by charlatans.

How to Find Civilizations around Other Stars

In the previous chapter, we adopted the posture that the Drake Equation provides a reasonable way of estimating the number of planets with life, and the number of civilizations, in the Milky Way. We shall postpone until the next chapter the discussion of how to recognize a planet with life. For now, let us recall that our best estimate of the terms in the Drake Equation implies that the number of civilizations in the Milky Way at any representative time equals about **L**/35, where **L** is the lifetime, measured in years, of an average civilization. If **L** lies somewhere between, say, ten thousand and ten million years, the Milky Way should contain thousands of civilizations, and the closest of them should be something like a thousand light years away. Where, then, should we look to find our neighbors? And what methods could we use to establish contact with them?

Interstellar Spaceships

The possibility that comes naturally to mind extends the human desire to explore the depths of interstellar space. Develop sufficiently powerful spacecraft, send them in all directions, and eventually these voyages of exploration will produce contact. This approach could work, but the time and expense would be more than astronomical. To begin with, our best spacecraft now travel at approximately one ten-thousandth of the speed of light, so using them to take a round-trip journey out to distances of a thousand light years and back again would take tens of millions of years. In addition, we have no way to supply these spacecraft with fuel for multimillion-year journeys. Furthermore, we likewise have no means of maintaining humans in space for many years, even if we could find teams willing to devote themselves and many succeeding generations to space exploration. Finally, we would need thousands upon thousands of these ships if we hoped to explore just the millions of stars that lie within six hundred light years of the sun.

These considerations make interstellar spaceships unlikely at least for the next few generations on Earth. Similar considerations should apply to other civilizations, though advances in technology could improve the situation—for example, by leading to spacecraft that travel thousands of times more rapidly than our fastest ones do, and which might find their fuel in interstellar space. A civilization confident of its survival might well launch explorers on journeys that last thousands or even millions of years. If they can create spacecraft that travel at nearly the speed of light, they can profit from an amazing aspect of Einstein's theory of special relativity: Time will appear to slow down for the space travelers, so that a journey that appears to require a million years, as measured on the home planet, would seem to take only a few years for the space traveler.

When Time Slows Down

Albert Einstein's theory of special relativity, first published in 1905, makes a startling prediction. If one object (for example, a spaceship) travels rapidly past another (for example, an observer on the planet Earth) then time will appear to slow down: the Earthbound observer will measure time as passing more slowly in the speeding spacecraft than it does on Earth! If the observer sees an object pass by at speed v, the factor by which time seems to run more slowly in the moving system equals one over the square root of the algebraic quantity $(1 - (v^2/c^2))$, where c is the speed of light. For velocities much less than the speed of light, this time-dilation factor exceeds one only modestly, but as v approaches c, the factor grows large. For $v = 0.995c$, the slowing-down of time reaches 10, and for $v = 0.99995c$, it climbs to 100.

Equally remarkable, consider the fact that the traveler in the spaceship will see the *Earth* appearing to move rapidly past the ship and will observe time on Earth passing more slowly than it does on the spaceship. How can this be so? If one twin embarks on an interstellar journey at nearly the speed of light and then returns to Earth, will this traveler or her stay-at-home twin be older when the twain meet? Physicists encountering the theory of relativity have thought hard about this paradox and have shown that Einstein's theory has consistent answers for all these questions.

First, we should recognize that our notions of time form another sector of the intuitive portion of our worldview. We cannot count on intuition, formed from experience in only a tiny fraction of the vast realm of the universe, to provide correct answers about anything as cosmic as the nature of time.

Second, once we have loosened our minds a bit, we can see, as Einstein did, that if two objects pass one another at high speed and never change their directions of motion, they can never get back together again for a side-by-side comparison of who has seen time pass more slowly. Any comparison must be made by exchanging messages, and these messages travel no faster than light. As a result, precisely because we cannot easily define the moment of "now" for two widely separated observers, each of them can picture the other, far-distant one to be aging more slowly without creating an absurdity.

Third, if we do arrange to bring the two observers together again, as we do with the paradox of the twins, then the two will not have identical experiences. We shall concentrate on the simplest situation, since life grows short: Imagine a twin sent to explore the planet around 51 Pegasi, 45 light years from Earth. Not only does the traveling twin accelerate and decelerate in her motion (this is an important detail, but it does not resolve the twin paradox). The key is that *the traveling twin reverses direction* to return along her original trajectory. To see what this means, let us follow a suggestion made by the Cornell University physicist David Mermin and imagine that each twin wears a beacon that flashes once every second, as measured by the twin who wears that beacon. While the twin travels outward at 99.5 percent of the speed of light, we might expect that each twin would see the other's time slow down by a factor of 10, so that the other twin's beacon would flash not once per second but once every 10 seconds. However, we must also consider how the traveling twin's motion affects the light-travel time for the flashes. Between each successive flash, the traveling twin moves farther from Earth by 9.95 light-seconds (9.95 times the distance that light travels in one second). Hence her flashes reach Earth not once every 10 seconds but once every 10 + 9.95 = 19.95 seconds. This takes into account both the slowing-down of time and the increase in the light-travel distance. The traveling twin likewise receives one flash from the twin on Earth every 19.95 seconds, so the twin paradox continues to apply.

Now consider what occurs when the traveling twin reverses direction to begin her homeward journey. On the return leg, the distance to Earth *decreases* by 9.95 light-seconds in the interval between successive flashes. Hence the flashes received from this leg will reach Earth at time intervals equal not to 10 seconds but rather to 10 − 9.95 = 0.05 second. Roughly speaking, once we take into account the changing distance, we can say that on the outward leg of the trip, the light flashes arrive at time intervals 20 times longer than they would if no motion occurred, whereas on the homeward leg, the flashes arrive 20 times more often than they would in the absence of motion.

But let us now ask when the Earthbound twin will observe the change from "20 times more slowly" to "20 times more rapidly." The answer is: not until the traveling twin has almost returned! Since she travels at 99.5 percent of the speed of light, she falls below light speed by only 0.5 percent of c. Long after she has reversed direction and started home, the pulses she emitted on her *outbound* leg are still arriving at Earth. The flash from her farthest distance, emitted at a distance of 45 light years from Earth, precedes her own arrival by only two and a half months. The Earthbound twin receives 20 flashes per second only during this last small fraction of the total time, since the traveling twin's flashes barely outrun her. In contrast, the traveling twin sees a dramatic change in pulse-arrival times *immediately upon reversing her direction of travel.* During the first half of her journey, pulses from Earth arrive once every 20 seconds, and during the second half, which lasts equally as long as *she* sees it, pulses arrive 20 times per second. The traveling twin thus has a far longer span (half of the total journey) during which she finds pulses arriving 20 times per second than the stay-at-home twin does. By following this example, we can keep track of all the pulses and verify that the traveling twin does age much less than the stay-at-home twin. For a trip made at 99.5 percent the speed of light, a round-trip journey of 90 light years will appear to take 90.45 years as measured on Earth, but the traveling twin will age by only a bit over 9 years. Here we have a way to let others age more than we do! At the price of losing all our friends and relatives, we could sail through space, not prolonging our life in our own frame of reference, but nevertheless stretching our experience over a wider band of time as measured on Earth.

Those who find this hard to follow, or to believe without understanding, should take note of the fact that Einstein's prediction has been repeatedly verified in particle accelerators that routinely shoot elementary particles in circular trajectories at more than 99.5 percent of the speed of light. Some of these particles decay, or change into other particle types, at a well-known rate—but when they travel at high velocities, they take longer to undergo these decays, as witnessed by the experimenters. The more their velocities approach the speed of light, the longer the particles take to decay, in just the relationship that Einstein had predicted in 1905, before a single one of these particle types had been identified.

Interstellar Radio and Television

Einstein's relativity theory shows that interstellar journeys requiring only a few years (as measured by those who make them!) are theoretically possible. Before we draw the conclusion that this would explain extraterrestrial visitors to Earth, we should note that *radio, television, and similar means of communication will, according to our present understanding of the cosmos, always offer cheaper and more rapid means of exchanging information than actual voyages can.* No civilization would be likely to ignore this basic truth about interstellar communication.

The superiority of radio, television, and other types of electromagnetic radiation lies in the fact that they travel at the speed of light, nearly 300,000 kilometers per second, and they cost remarkably little to produce. We can measure "cost" in energy terms and note that the Earth now contains thousands of radio and television stations, each of them broadcasting in all directions 24 hours per day, at an ongoing energy cost not much greater than needed to run a large diesel truck. In comparison, the *Saturn V* rocket that carried astronauts to the moon cost several billion dollars to develop, a billion dollars to build, and consumed enough fuel to run 50,000 diesel trucks for a week. This fuel was necessary to accelerate the rocket to the modest velocities described above (less than 1/10,000 the speed of light), to decelerate the spacecraft and set it into orbit around the moon, to accelerate away from the moon, and to decelerate again near Earth. Any round-trip journey requires these four accelerations or decelerations, not simply the first acceleration that takes the spacecraft away from Earth and gives it a large velocity through space. We must never forget that the closest stars have distances tens of millions of times greater than the distance to the moon.

In contrast, electromagnetic radiation, once generated, travels through space at the speed of light until something blocks its passage. A round-trip radio or television message exchanged over a distance of 1,000 light years requires 2,000 years for the trip—not a brief time, but one that humans can understand and even plan, if we choose to, with our present capabilities. Should we broadcast radio and television messages saying, "Here we are"? In a sense, we already have: The radio, television, and radar emissions from Earth have spread outward in all directions, in ever-growing amounts. By now, the emissions produced during the

Second World War have reached distances nearly 60 light years from the solar system, passing by hundreds of stars in the process. A civilization around any of the few hundred nearest stars might have detected these emissions, all jumbled together, and have recognized that they come from a rotating planet on which the transmitters are clumped into certain areas. This might suggest that Earth has oceans and dry land, since indeed that fact is largely responsible for the concentrations of transmitters. The oft-repeated suggestion that another civilization might have received our television broadcasts, watched a few episodes of old sitcoms, and rejected us as fit for intelligent conversations seems highly unlikely, if only because separating the totality of Earth's emission into different programs appears a nearly impossible task to us.

However, we have been more direct in announcing our existence. In 1974 astronomers broadcast a radio message from the Arecibo Observatory in Puerto Rico. This message consisted of 1,679 bits of information, basically either an "on" or an "off" pulse. Another civilization might receive the message, realize that 1,679 is the product of two and only two prime numbers, 23 and 73, and try arranging the bits into 23 rows of 73 columns. If the "on" and "off" bits are then colored differently, a pattern emerges. Another civilization might recognize this pattern as nonrandom and might even deduce which parts of the pattern represent the Earth, the humans that sent the message, and the radio dish they used. The Arecibo message of 1974 was sent toward the globular star cluster M13, which lies 25,000 light years away, so we cannot expect any reply for nearly 50,000 years. We could, however, choose to direct such messages toward much closer stars. And, of course, we could turn our radio telescopes toward the closest stars, to see whether they might be broadcasting "Here we are" signals or whether we might detect their overall radio and television emission, if they use these means for their own communications.

Why do we focus on radio and television signals? The radio region of the electromagnetic spectrum has the advantage of offering the least expensive way (in energy terms) to send messages; light pulses cost considerably more, because they each have much higher energy than radio waves do. More important, the Milky Way is relatively quiet in radio waves but sparkles with starlight in the visible, ultraviolet, and infrared regions of the spectrum, so any signals meant to be detected face less competition from distracting noise in the radio than in higher-frequency regions of the spectrum. Interstellar background noise becomes significant at the lowest frequencies, which gives the frequencies around 1,000 megahertz the best chance to carry a signal that can be detected against the noise background.

In 1992 the United States Congress funded a project to search the cosmos for possible extraterrestrial radio signals, then promptly reversed itself, so that funding was eliminated in 1993. Nothing daunted, the astronomers interested in SETI (Search for Extraterrestrial Intelligence) raised sufficient private funds to promote Project Phoenix, an ongoing search for extraterrestrial radio signals that might arise from other civilizations. So far, the project has found hundreds of signals that have clearly non-natural origins—but they all turned out to arise on Earth. The same is true for Project Serendip, which piggybacks on observations made by the Arecibo radio telescope: Whenever the telescope observes an object, a small part of the signal enters the Serendip detectors, which analyze it for possible signs of a civilization-made broadcast—so far without success. Although Project Serendip does not choose its targets, over the course of years it surveys much of the sky observable with the Arecibo telescope, simply because radio astronomers observe a variegated range of sources. Serendip's observations include four of the stars listed in Table 1.1 (which, we should recall, do not seem to present particularly favorable sites for life to exist).

This might seem to imply that the Milky Way contains few civilizations, or at any rate few that use radio and television broadcasts in anything similar to the manner that we do. But we should note that the searches involved in Project Phoenix, and in other more modest search attempts, face a formidable obstacle:

the enormous possible spectrum of radio frequencies and wavelengths to be searched. These range from the short-wavelength radio waves called microwaves, with wavelengths measured in millimeters, to radio waves 100,000 times longer, with frequencies only 1/100,000 as large. In contrast, visible-light wavelengths and frequencies vary by only a factor of two or three. Even if astronomers restrict their search to radio frequencies, they face an additional problem: How large a bandpass should they use? The bandpass measures the span of frequencies under observation. But if the bandpass far exceeds the span of frequencies that carry a signal, the observation wastes most of its effort, since it mostly observes frequencies that have no signal. On the other hand, if the observations use a bandpass much narrower than that of the signal, it likewise wastes effort, because it misses most of the signal.

Hence astronomers who search for signals from other civilizations face a threefold problem: Where should they look? Which frequency carries the signal? What bandpass should they employ in their search? The first question has a reasonable answer: Search among the few hundred closest stars and also make a general search of all directions in the Milky Way, to the extent that time and budget permit. To the second question, the answer has emerged: Search millions of frequencies simultaneously! Amazingly, systems now exist that can do just this, as if a radio could listen to millions of stations at once. Even so, these searches cannot cover the full spectrum of possibilities, but they offer a good start at doing so. And to the third question, no good answer exists, but astronomers use as narrow a bandpass as seems compatible with completing a reasonable search in a reasonable amount of time.

So when you hear that astronomers have found no signals from extraterrestrial civilizations, despite years of effort, remember that in terms of sweeping the entire sky, over literally billions of possible frequencies, with an unknown ideal bandpass, you may be wise to conclude that we shall need far greater efforts to find our closest neighbors. They may in fact not exist: We might be alone in the Milky Way, if not in large regions of the cosmos. If this proves true, at least we are starting to calibrate our importance as a civilization by determining the difficulty of finding others something like ourselves. If it does not, we shall someday look back on the 1990s, from the perspective of success in achieving contact with other civilizations, and say, *That* was the time when serious searching for other civilizations first began.

Meanwhile, the more modest quest for extrasolar planets continues. Having surveyed the results to date, we owe it to ourselves to leave the subject with a look at plans for the future: to search for extrasolar planets from Earth-based observatories; from telescopes in Earth orbit; and from the mightiest system of all (counting only astronomers' current dream projects), from an interferometer sent to Jupiter's distance from the sun.

CHAPTER

9

Future Searches for Extrasolar Planets

The discovery of planets around other stars has led to a host of activity designed to survey and categorize our neighbor worlds in the Milky Way. In a brief span of time, searches for extrasolar planets have passed from a quiet backwater to join the mainstreams of astronomical research. Though some astronomers resent the notion that precious resources might be diverted from cosmology, stellar evolution, or other astronomical research to hunt for new planets, most of the astronomical community has welcomed what amounts to a new subject—actual planets around other stars—both for its implications about the cosmos and for its ability to engage the public's interest.

Nowhere has the news about planets made a greater impact than at the National Aeronautics and Space Administration. In 1995 NASA announced that its space-science research will focus on four fundamental issues: sun-Earth relationships, the exploration of the solar system, the search for origins (of galaxies, stars, planets, and life itself) and other planetary systems, and the quest to understand the structure and evolution of the universe. By creating the category of "origins," NASA united astronomers' desires to understand the full history of the universe with the public's urge to connect with the cosmos. This connection arises most straightforwardly in contemplating the universe as a whole, with its spiritual overtones, and in considering the possibilities of life on worlds other than ours.

Given NASA's recent emphasis on "origins" and the multitude of new planets discovered in 1996, we may expect that the end of this millennium, and the start of the next, will find United States space research emphasizing the search for planets, even though that search will receive only a small fraction of the total NASA budget. The fourteen European governments who collectively fund ESA, the European Space Agency, do not look to public opinion so openly as NASA does; nevertheless, those who direct their governments' support for scientific research are quite aware of the public interest in exploring for planets and will add this fact to their deliberations about the projects they choose for the first quarter of the next century. Scientists in both the United States and Europe naturally hope to combine their efforts in the most important research projects, thus avoiding wasteful duplication and assuring that the chosen projects have the best chance of success.

What lies in the future—say, the next quarter century—in searching for extrasolar planets and understanding their nature? We can divide the increased efforts into three categories, based on the location (and thus, to a large extent, the expense and difficulty) of the observational programs. First, ground-based astronomers will design, create, and operate new telescopes and new interferometer systems on Earth, capable of producing single-point images of giant planets around stars within a few dozen light years of the sun. They will also improve their Doppler-shift, astrometric, and microlensing observing programs, to achieve greater planet-detection capabilities and to refine their studies of planets already found. Second, astronomers oriented toward space-borne observing platforms hope to send into orbit the Next Generation Space Telescope, which should see giant planets more easily, though it probably cannot hope to find Earth-like planets around even the closest stars; they may also create an orbiting telescope to search for planets that pass directly in front of their stars. Third, at the far edge of the twenty-five-year horizon, visionary astronomers dream of constructing an automated interferometric observatory at distances of 4 to 5 A.U. from the sun, comparable to Jupiter's distance, to search for Earthlike planets around the few thousand closest stars. Only by sending such an observatory far from the sun can they overcome the interference from a pesky problem, called the zodiacal light, in their efforts to win a prize in the search-for-an-Earthlike-planet lottery. By examining each of the three broad categories, we can win our own prize, an understanding of the possibilities and limitations on what can be done in the hunt for planets.

The Road Map: A U.S. Guide to the Search for Planets

During the past decade, NASA has received large doses of opprobrium, much of it deserved. Amid NASA's woes and mistakes, analysis of how to search for extrasolar planets stands out (so far!) as a well-organized, forward-looking undertaking. Early in 1995, months before the announcement of the first extrasolar planet, NASA asked researchers at many institutions to create a road map of the journey toward the discovery of Earth-like planets around other stars. The researchers created three study groups to engage in friendly competition and to present their ideas to an "integration team" at the Jet Propulsion Laboratory in Pasadena, which combined the best of them into a road map for review by a panel of distinguished scientists led by Charles Townes, the Nobel Prize–winning physicist and astrophysicist. The review panel in turn made its recommendations to NASA. Because these recommendations arrived in early 1996, while a series of budgetary and political crises temporarily paralyzed the government, they received relatively little publicity. The road map recommended by the review panel now serves as the basic guide to the Exploration of Neighboring Planetary Systems program (ExNPS in NASA acronymistics, pronounced *ex-nips*), which covers all efforts to find extrasolar planets that the federal government may support.

The review panel concluded that "[t]here exists perhaps no foreseeable discovery that would more electrify the public imagination and spark a renaissance in science education and science literacy than the direct detection and characterization of earthlike planets around the closest 1,000 stars. The potential unifying impact on astronomy, planetary science, and geophysics would be equally profound. . . . [A] unique opportunity exists for a great civilization to attack and to solve, on a scientifically sound and technologically feasible basis, one of humanity's oldest and deepest questions concerning physical origins. . . . We urge NASA and the nation not to let this opportunity pass." In the next sentence, the panel noted that since ExNPS is ambitious both technically and scientifically, it should not become dependent on "too many technically-challenging elements which may or may not develop as hoped."

On the other hand (for the government has many), the review panel concluded that "the goal of understanding other planetary systems can be pursued more advantageously in space with devices that provide high spatial resolution" and stated that the panel's members had been "impressed with the strong case that can be made for the utility of a cooled interferometer that is sensitive in the infrared and orbiting at [a distance of about] 4–5 A.U. from the sun. The greatest advances in our knowledge of other planetary systems may well come from such an instrument." Since this instrument, described later in this chapter, represents the culmination of astronomers' current plans for finding planets, the panel recognized that although this interferometer "currently appears to be the natural and best choice for achieving the ultimate ExNPS objectives, there are nevertheless a number of scientific and technical hurdles to pass before making a large commitment to this ambitious mission."

Let us, like NASA's review panel, consider the various steps toward a system that might find and characterize Earth-like planets, starting with the proven techniques that have already revealed planets and proceeding to reach the astronomers' grand design of an interferometer near Jupiter. As occurs in the real world, we shall encounter various decision points, rife with discussion among proponents of different approaches toward achieving the most accurate vision of the cosmos at the lowest cost. Since we must eventually pay for both the roads and the motels on the road map, we owe it to ourselves from our perspective as taxpayers to become knowledgeable about the various possibilities. Then, when our family and friends ask which road leads to other Earths, we can provide our conclusions to aid them in their own deliberations.

How Far Can We Go with Doppler Shifts and Astrometry?

Astronomers discovered significant numbers of new planets during 1996 by using the time-tested, though much improved, Doppler-effect technique: by making precise measurements of the wavelengths of the absorption lines present in the spectrum of starlight. They also began to see the results of astrometric programs, which make detailed observations of stars, capable of revealing deviations from straight lines in the stars' proper motions. During the next decade, we can expect several types of improvements in these two observational techniques, which should significantly increase the number of detected extrasolar planets.

First and most obviously, success brings its rewards. Observational astronomers' rewards consist of more time granted to use the world's largest telescopes and greater funding to improve their detector systems and to process their data. These advantages are now accruing to Marcy and Butler in California, to Mayor and Queloz in France and Switzerland, and to others who have devoted a large part of their careers to searching for planets around other stars. Furthermore, Doppler-shift and astrometric searches improve with time, even if the accuracy of a given observation remains constant, because the observations yield a longer string of data to be analyzed. This leads to greater accuracy in finding a cyclical pattern in the data. Within a few years, the planet searchers in the United States and Europe will have at their disposal a much larger number of observations, spread over significantly longer spans of time, than they did only a few years ago. Unless fate is playing a cruel trick, this should lead to a sizable increase in the rate of discovering new planets.

For the Doppler-shift searches, the net effect of improved instrumentation, greater allocations of observing time, and longer time periods over which data has been collected will be to push down the line in Color Plate 22 that divides the upper region, within which planets can be detected by this method, from the lower region, where they cannot. The astrometric searches are relatively less advanced, so that a similar diagram cannot be readily drawn, but in an analogous diagram for this method, the "accessible region" would fill the upper right-hand side, including all distances greater than a few A.U. and extending downward to planetary masses perhaps 10 percent of Jupiter's mass. Of course, this presupposes observations made over a sufficient time span—many years at least!—to reveal the changes in a star's proper motion produced by a planet in a multiyear orbit.

What about Gravitational Microlensing?

In chapter 6, we saw that gravitational microlensing offers a means of detecting planets around other stars out to distances hundreds or even thousands of times greater than any planets found by the Doppler-shift or astronometric methods. In presenting its road map, the NASA review panel considered this technique and gave it only modest encouragement, stating that "the Panel encourages the microlensing community to continue with their observational program. But other ground-based approaches to detecting planetary systems in the solar neighborhood such as occultation, adaptive optics, spectroscopic [Doppler-shift], and interferometric methods also need attention and development." From an ExNPS perspective, gravitational microlensing cannot say much about the solar neighborhood; instead, this approach will reveal the statistical distribution of planets at much greater distances from the sun than any nearby star. If we seek to characterize worlds in our own corner of the Milky Way, microlensing can't do much, but it offers our best chance to learn something about the numbers of planets, and their distances from their stars, throughout our galaxy.

The Next Generation Space Telescope

Despite the impressive advances soon to occur in Earth-based observations of the cosmos, sending telescopes into space will always represent a local improvement in our observational capacity, because some of the limits set by our atmosphere can never be overcome with better technology. For example, Earth's veil of air blocks almost all gamma-ray, X-ray, and ultraviolet radiation, along with large portions of the infrared and microwave spectrum. Since life on Earth evolved with this blockage in place, we should not be surprised to learn that much of this radiation will destroy almost all forms of life; hence we tamper at our peril with the atmosphere's ability to absorb ultra-violet radiation. So long as we engage in no such tampering, our atmospheric shield will remain in place. In that case, if we seek to observe the cosmos in ultraviolet, X rays, gamma rays, or certain infrared wavelengths loaded with information about the universe, we must send instruments beyond our protective veil of air.

NASA did just this with the launch of the Hubble Space Telescope (HST) in 1990. The HST's ability to study the cosmos over a range of ultraviolet wavelengths completely blocked by our atmosphere has proven nearly as useful to astronomers as its ability to observe with an angular resolution better than any Earth-based telescope can achieve. As we discussed in chapter 3, however, even the HST cannot hope to see planets around nearby stars. Though the 1999 mission to install an advanced camera system may actually provide the Hubble Space Telescope with a marginal ability to see nearby extrasolar planets, the great hope for space-borne astronomy now resides with the Next Generation Space Telescope (NGST). The NGST would concentrate on infrared astronomy, especially important in studying young galaxies billions of light years from Earth, and in searching for protostellar disks and possibly for planets as well. In contrast to the 2.4-meter mirror in the HST, the NGST should have a mirror between 4 and 8 meters in diameter, giving it far better angular resolution than its predecessor, which will not be renamed the Last Generation Space Telescope. Equally important, the NGST will have technology of the new millennium, not that of the 1970s, when the HST was built. Since the HST's problems drove an already high cost toward exorbitance, NASA supports the NGST only if its cost can be held to $500 million, about a quarter of what the Hubble Space Telescope cost.

Let's see—a telescope whose mirror has two or three times the diameter, hence four to nine times the area of the HST, but must cost only one quarter as much, in an inflationary situation when all of science faces much greater competition for the budget dollar. How do they plan to do that? The main line of attack consists of finding radically new ways to build a large mirror. The cost of launching objects into space depends crucially on their masses, so the astronomers must find a way to make a much larger mirror that weighs much less. John Mather, the chief study scientist for the NGST at NASA's Goddard Space Flight Center, believes that a mirror only a few millimeters thick can be designated, fabricated, launched, and deployed in a telescope that would be sent into an orbit, not a few hundred miles above the Earth, like the HST's, but instead 93 million miles from Earth, to one of two possible locations, each at the Earth's distance from the sun but always 60 degrees ahead of or behind the Earth in its orbit. At either of these two Lagrangian points, the NGST would orbit around the sun without being slowly moved into a new orbit by the gravitational forces from the Earth and the sun.

In its present orbit, the HST loses large amounts of time and flexibility from always having an immensely bright object (the Earth) only a few hundred kilometers away; to point the telescope in that direction for even a second would destroy the sensitive optical system. An orbit at one of Earth's Lagrangian points would minimize this problem. Furthermore, far from the radiation reflected by the Earth, the NGST would cool itself as the telescope radiates heat into space. The temperature would fall to a few tens of degrees above absolute zero, a great asset in making infrared observations: At significantly higher temperatures, the telescope itself radiates far more infrared, interfering with any attempt to observe radiation from far-distant reaches of the cosmos.

Mather thinks that the NGST can be built for $500 million; other astronomers say they would be flabbergasted, though delighted, if this can be accomplished. One thing is clear: To achieve this feat, NASA's scientists will have to find some new ways to make large mirrors. "We've got to get off the kick of extrapolating ground-based telescope designs into space," Dan Goldin has said. Finding a way to build and to launch a thin mirror, one that will become a central part of this effort.

Since the NGST design studies have barely begun, any pronouncement about its ability to detect extrasolar planets seems premature. However, we may note that although the NGST might prove immensely useful for seeing planets—as individual points of emission, not in any greater detail—it will not attempt to find planets by the transit method described in chapter 6. That method requires a dedicated instrument, one that can spend essentially all its working hours checking the brightnesses of stars in search of small changes arising from the transit of a star across the stellar disk. In short, we can't hope to get a "twofer" in planet searches with the NGST. This provides no argument at all, however, against the NGST, which promises to be the greatest astronomical instrument ever built. Only the decade of brilliant innovation required to build and to launch the NGST at remarkably low cost stands between the present epoch and a glorious future at the end of the next decade, when the Hubble Space Telescope (which will still be operating, with an upgraded camera system, if Congress appropriates the funds) watches its successor, the NGST, pass by to enter a far more useful orbit than the HST's, a half a million times farther from Earth.

Better Earth-based Observations: Adaptive Optics and Interferometers

Throughout human attempts to observe the universe run the difficulties and restrictions imposed by our atmosphere, life-giving yet never perfectly transparent nor quiescent, which absorbs most cosmic radiation other than visible light, radio waves, and some wavelengths of infrared radiation. Earth-bound astronomers have struggled for centuries to peer more effectively through the curtain of air toward the heavens above. They have created the world's finest telescopes, sited at high-altitude observatories with remarkably clear and still skies, and have fitted these telescopes with the most sensitive equipment they can construct to detect and to analyze the radiation captured by these instruments. By now, they appear to have reached the limits set by the constantly rippling layers of air above locations such as Mauna Kea, Hawaii, and Cerro Tololo, Chile. What more can be done on Earth?

The answer to this question proves as amazing as it is subtle. We have stated as a bald fact that the constant motion of the atmosphere causes starlight to bend by an ever-changing amount, and that this varying refraction of light prevents astronomers from observing any object through the atmosphere with an angular resolution better than a few tenths of a second of arc. This is true—if you attempt to vanquish atmospheric refraction by brute force, by simply constructing a fine telescope at the best observatory site you can find. But what if astronomers found a way to monitor the atmosphere continuously, following its changing refraction and then compensating for it with an optical system that can change its properties slightly, responding in a tiny fraction of a second to track the changing refraction? Quite a mouthful—and quite a challenge, but one that astronomers have now risen to meet. They have developed several different systems of adaptive optics, each designed to produce slight changes in the optical properties of a telescope that will compensate exactly for the small atmospheric changes that otherwise will spread any image over at least a few tenths of an arc-second.

Most prominent among the proponents of adaptive optics is Roger Angel, an English-born astronomer at the University of Arizona and an expert at building large, thin telescope mirrors that can be continuously adjusted to overcome the effects of the changing refraction that produces atmospheric "seeing." Angel hopes to construct a 16-meter reflector with adaptive optics, capable of obtaining images of Uranus-sized planets at distances of at least 5–10 A.U. from nearby stars. This instrument could, for example, see the planets that George Gatewood has claimed to have found in orbit around Lalande 21185—if they exist. Angel, a man whose dreams live up to his name (in 1996 he received a "genius award" from the MacArthur Foundation) claims that a 32-meter adaptive-optics reflector could even produce images of *Earth-like* planets around other stars. Other astronomers dispute this claim, arguing that Angel has been too optimistic about the performance of his proposed telescopes. Nothing daunted, Angel sometimes counters these arguments with a proposal for an Earth-based interferometer system with four 16-meter telescopes, each equipped with adaptive optics.

How can we assess these proposals? Even an expensive Earth-based telescope tends to be far *less* expensive than a telescope in space, which must have a guidance system that keeps the telescope pointed accurately toward one spot in the cosmos as it orbits at 27,000 kilometers per hour, along with systems that collect, store, and transmit data. Almost everyone agrees that adaptive optics has a bright future, and that by building the first large telescope with this ability, we shall make impressive strides forward with Earth-based observatories, and also create a test bed for future, larger telescopes. After all, the world's largest telescopes, the twin Kecks of Mauna Kea, have mirrors "only" 10 meters across. The Kecks have active optics (note the danger of confusion with adaptive optics), meaning that their mirrors, which actually each consist of 36 hexagonal panels, constantly re-align themselves to compensate for the changing gravitational stresses that arise as the telescope moves to different positions. The Kecks' active optics have performed entirely up to design specifications, but these never included adaptive optics, the attempt to compensate for constant changes in atmospheric refraction. Stay tuned for news from Roger Angel and his competitors, because adaptive-optics telescopes are the wave of the future for ground-based astronomical observations in visible-light wavelengths. To find Earth-like planets, however, many astronomers believe that we must go above the atmosphere.

This will not curb Angel's enthusiasm, nor that of the astronomers who are developing optical and infrared interferometers for Earth-based observations. As we described in chapter 3, an interferometer combines the radiation detected by two or more telescopes to produce a single image with the angular resolution of a single telescope whose diameter equals the maximum distance between the individual telescopes in the interferometer. Since image-oriented planet searches aim to see a dim planet next to a bright star, the key question becomes, What kind of interferometer can achieve this goal for various planetary configurations? The answer to this question seems to converge on a system sent out almost to Jupiter's distance from the sun.

Mission by Jupiter: The Quest to Find Other Earths

The search for extrasolar planets will be seen as an important part of the human experience once astronomers discover other Earths and assess their suitability for life. As we have seen, the newly discovered planets, most of which have masses similar to Jupiter's but orbit their stars at distances far less than the distance between the sun and Mercury, seem unlikely to contain other

Earths for possible detection. In order to find and to characterize extrasolar terrestrial planets, it appears that we must build an interferometer system for infrared observations, capable not only of securing images of distant objects but also of making detailed observations of these objects' spectra; we must send the system out to almost 4 or 5 A.U. from the sun; we must cool the detectors to low temperatures; we must arrange for the system to maintain its precision to a fraction of a wavelength's distance; and we must have the ability to direct the system's observational program and to receive its data for detailed analysis.

When I say "we," I mean, of course, other people. An obvious aspect of the system summarized above will be its complexity, and the average person's role will be to help pay for the effort in order to savor its fruits. One of the key unanswered questions concerning this project—one that may prove more difficult to resolve than any technological or scientific issue—is whether it will be a United States or a worldwide effort. The future holds the answer, but describing the vision does not by itself require global integration. We need an interferometer system because only an interferometer can provide the angular resolution needed to show a planet as an object distinct from its star. In order for this interferometer to function properly, it must maintain the distances between its components to a constancy of a few percent of the wavelength at which it observes—better than 1/10,000 of a millimeter. The interferometer must observe in the infrared because in that spectral region, a star outshines an Earth-like planet by "only" a factor of about a million, rather than by a billion times, as occurs in visible light. Because the system tends to emit its own infrared radiation that will lessen its ability to detect faraway objects, we must cool the system to reduce this problem. We must send the interferometer system out to 4 or 5 A.U. from the sun because the *zodiacal light*, sunlight reflected by dust, shines so brightly in the inner parts of the solar system that the type of planetary detection we seek to achieve becomes impossible from there. We require fine spectral observations to determine the surface and atmospheric characteristics of any planet we detect, all of which affect the details of how the planet reflects and emits radiation. And we must have the basic ability to direct the system's observing program by sending radio signals, as well as the ability to receive signals encoding what the interferometer detects.

What Is the Zodiacal Light?

The reader may take it as a happy omen that he or she understands these requirements—all, perhaps, except the mysterious "zodiacal light." The name emphasizes that this faint glow, visible as dusk ends or dawn begins, concentrates along the plane of the zodiac, the region containing the orbits of the sun's planets. In 1661 the British naturalist Joshua Childrey recorded the first accurate description of the zodiacal light, stating that "In February . . . when the Twilight hath almost deserted the horizon, you shal see a plainly discernable way of the Twilight striking up towards the Pleiades or Seven Starrs, and seeming almost to touch them." Today, thanks to the progress of civilization, we can see the zodiacal light only under favorable dark-sky conditions, far from any city, just before sunrise or just after sunset—but probably almost never from the regions in England where Childrey lived.

On the other hand, we now understand what produces the zodiacal light: dust particles throughout interplanetary space, which reflect sunlight and also emit infrared radiation. This interplanetary dust recalls the long-vanished eras when the sun's protoplanetary disk provided a large mass of gas and dust from which planetesimals could form. Today, relatively tiny amounts of dust orbit the sun, generated by collisions among objects in the asteroid belt, but even this dust creates a problem. Whenever astronomers seek to observe faint objects, they must do so amid the faint but constant glow from the zodiacal light. Those familiar with the problem often shorten the name to the "zodi light" or the "zodi cloud," a reference to the fact that the zodiacal light arises from a solar-system-wide cloud of dust. Even though the zodi light concentrates along the plane of the zodiac, we on Earth are sufficiently deep within the zodi cloud that its light interferes with faint-object studies in all directions.

Since the quest to see extrasolar planets raises faint-object observations to a new plane of difficulty, the zodiacal light rises to prominence—negative prominence, to be sure—in all these efforts. The dust that reflects sunlight and emits its own infrared fogs our view, so that an attempt to observe the faintest objects resembles an attempt to see stars in daylight, when the background glow from the sky (the result of sunlight reflected by molecules and dust particles, in this case high in Earth's atmosphere) interferes with efforts that succeed easily at night. The zodiacal light knows no nighttime: Even if we built and launched an infrared interferometer system theoretically capable of revealing Earth-like planets, to attempt to use it close to our planet would prove fruitless. Our only hope for avoiding the "bright-sky" conditions from the zodi cloud lies in journeys farther from the sun. From measurements made by the *Pioneer* and *Voyager* spacecraft, we know that the amount of interplanetary dust decreases steadily as we move outward in the solar system. The amount of sunlight to heat the dust and to reflect from it also decreases, aiding our attempts to escape the zodi glow. By the time we approach Jupiter's distance, the infrared emission from the zodi cloud has declined to the point that an interferometer system can perform its appointed task of searching effectively for other Earths.

Interferometers of the Future

What kind of an interferometer system will then be required? How can scientists design and build such an instrument, meant to operate a billion kilometers from Earth while maintaining a fantastically high precision in the alignment of its components? The only obvious segment of the answers to these questions deals with the incremental advances that technological progress requires. In order to create an infrared interferometer able to find extrasolar Earths, scientists must first build interferometers on Earth, then test interferometer and infrared systems in space, and finally produce the system to be sent past the bulk of the zodi cloud to achieve the success to which they aspire.

The late 1990s will see significant improvements in Earth-based interferometry as five new interferometers reach completion. These include two systems in Arizona, the Navy Prototype Optical Interferometer (*NPOI*) at Lowell Observatory and the Infrared-Optical Telescope Array (*IOTA*) at Mount Hopkins Observatory, as well as the Cambridge Optical Aperture Synthesis Telescope (*COAST*) in England, the Sydney University Stellar Interferometer (*SUSI*) in Australia, and the Astronomical Studies of Extrasolar Planetary Systems (*ASEPS*) interferometer at the Palomar Observatory in California. All five of these interferometer systems rely on the same fundamental principles in their operation, though they differ in the details of how they compare the light collected by two or more instruments. The *ASEPS* system, directed by Michael Shao of the Jet Propulsion Laboratory in Pasadena, represents NASA's entry in the interferometer sweepstakes and will be used to search for planets by the astrometric method. To do this, *ASEPS* will observe two stars in a single field of view, using one star as reference object from which to measure changes in the other star's position.

As these interferometers yield positive results, astronomers plan to create new and improved Earth-based interferometer systems. The most ambitious of these will be created at the Mauna Kea Observatory in Hawaii. Once the *ASEPS* system has been "proved out" at the Palomar Observatory, its designers plan to build a similar but larger system that will include the twin 10-meter Keck Telescopes on Mauna Kea, along with six 1.5-meter telescopes to be installed nearby. A similar plan exists to construct the Very Large Telescope Interferometer (*VLTI*) at Paranal, Chile, using the four 8-meter reflecting telescopes to be built there, along with eight 1.8-meter telescopes.

These ambitious projects testify to astronomers' recognition of the power of interferometry, from which radio astronomers have long benefited. Because light waves have much shorter wavelengths than radio, building a visible-light interferometer has proven far more difficult than once anticipated. The construction of an interferometer system for short-wavelength infrared radiation, which can penetrate the Earth's atmosphere (at least to higher elevations), has been additionally hampered by the need to develop more efficient detectors of infrared. Once improved infrared detectors reach the astronomical market, astronomers will be able to profit from the fact that infrared radiation has wavelengths several times greater than those of visible light, though far shorter than those of radio waves. This makes an interferometer system operating in the infrared easier to construct than a visible-light interferometer. In deploying these future interferometers, astronomers hope to fit their new telescopes with adaptive optics. Then they will have a double winner: interferometers that combine individual images that each have better angular resolution than any obtained before.

The new generation of Earth-based interferometers should obtain images of Jupiter-sized and even Uranus-sized planets around nearby stars, even if the planets orbit at distances similar to Jupiter's and Uranus's distances from the sun. They will also provide astronomers with essential experience to build interferometers in space, along with the infrared detectors that will make them capable of finding relatively small planets. Infrared detector technology has also been improving rapidly, first with the Infrared Satellite Observatory (*ISO*), launched by the European Space Agency (ESA) in 1995, and soon with the Space InfraRed Telescope Facility (*SIRTF*), which NASA hopes to launch in 2001. *ISO* now orbits the Earth in an elliptical trajectory that keeps it 1,000 to 70,000 kilometers above the Earth's surface. To lower its own infrared emission, *ISO* uses liquid helium to cool its 0.6-meter mirror and its instruments; the slow evaporation of the two tons of helium carried into orbit will limit *ISO*'s useful lifetime to about two years. *SIRTF* will carry a 0.8-meter mirror and a camera system more advanced than *ISO*'s into a much larger orbit, thus avoiding interference from the Earth's infrared emission; its effective lifetime will likewise be limited by the evaporation of its helium coolant. The Japanese space agency also plans to launch another infrared satellite, *IRIS*, in the year 2000, so the future of space-borne infrared observations appears rosy.

To be sure, neither *ISO* nor *SIRTF* nor *IRIS* has interferometric capability. Astronomers cannot plan to construct space-borne interferometers until they use their Earth-based systems to develop the technology needed for exact metrication, the measurement of the distances between an interferometer's components to a fraction of a thousandth of a millimeter. On Earth, solid ground provides a stable platform for maintaining and measuring a particular separation between the components; in space, we must either provide a platform tens of meters long or else develop a means of continuously adjusting the distance between individual orbiting components to a precision undreamt of until recently. The former solution threatens to cost too much, because it requires strong, massive beams to be launched out to immense distances from Earth; the latter approach holds more promise, though for now we lack the technology to make it happen.

Nevertheless, astronomers hope to see a space-borne interferometer system, sent beyond the worst of the zodiacal cloud and capable of observing extrasolar Earths, by the early 2020s (Color Plate 24). (How far away that seems! Yet I recall that in 1955, *Fortune* magazine had a cover story about the year 1980, then 25 years away. The thought stunned me—but the year came and went without fulfilling all of the editors' predictions.) So far, no one at NASA has chosen a name for this project, which might be called Find Earths And Scrutinize Them (*FEAST*). Quite probably, the crucial question governing *FEAST*'s feasibility will turn out to be whether or not NASA and ESA can create a collaboration that will allow them to build and to launch a single mission, or whether, for complex political and technological reasons, the two chief centers of space exploration adopt separate trajectories. In the latter case, even a modest imagination leads to the conclusion that both projects may prove too expensive for their respective governments to support. Let us put aside this unhappy

possibility and consider ESA's project from the perspective that drives both ESA and NASA toward the search for extrasolar Earths: How do we know life when we see it?

When We Find Other Earths, How Will We Recognize Life?

Like many of us, NASA administrator Dan Goldin once had the dream of examining "Landsat photographs" of Earth-like planets around other stars. The optical facts of life imply that that will not happen in our lifetimes, since these images would require sending not a "simple" interferometer system out to Jupiter's distance from the sun, but rather a mammoth battery of telescopes, spaced many kilometers apart and capable of maintaining that spacing to a millionth of a millimeter. Our descendants may well compare photographs of Earth's cousins, but we shall not. We can adjust our goals, as Dan Goldin did, and dream of an almost equally appealing prospect, assessing Earth-like planets as to their suitability for life, and even for the presence or absence of life.

"The goal is life," says Alain Léger, an astronomer at the Institut d'Astrophysique Spatiale of the Orsay branch of the University of Paris. Léger has become a leader in the attempt to persuade ESA to choose the search for extrasolar planets, and for life around these planets, as the key project of ESA's Horizon 2000, a statement of its goals for the new millennium. "The key date in planning this search was 1980," Léger says, "when Toby Owen suggested that the key criterion indicating life is the massive presence of oxygen." Tobias Owen, a planetary expert at the University of Hawaii, has noted that oxygen molecules combine so readily with other types of atoms and molecules that even though a modest amount of oxygen in a planet's atmosphere can be explained by nonliving processes, whenever and wherever we find a large abundance of oxygen, we have encountered a high probability of life.

Certainly the history of Earth tends to support Owen's assertion. During the first two billion years after Earth formed, our atmosphere contained only modest amounts of oxygen, no more than a few percent of the total and probably less. At some point between about 2 and 2.5 billion years ago, however, immense numbers of tiny organisms floating in the oceans sent large amounts of oxygen into the atmosphere as part of their metabolic processes. This raised the fractional abundance of oxygen past 10 percent, and eventually close to the 23 percent we find today. To many living creatures, that oxygen was the most noxious pollutant ever released on Earth. Many of them succumbed to this poison; many others managed to find refuge from oxygen and have survived as the anaerobic bacteria we find today. Still other organisms evolved defenses against oxygen pollution, and yet others evolved to take advantage of the new abundance of oxygen. Precisely because oxygen combines easily with other molecules in reactions that release energy, oxygen represented a new, widespread source of energy, creating an evolutionary selection in favor of organisms that could draw energy from the oxidation process. Today, every breath we take testifies to our ancestors' successful evolution to use oxygen in their metabolism. At bottom, respiration resembles rusting (slow oxidation): Both cause oxygen in the air to combine with other molecules in processes that release energy. All around the globe, rust-colored rocks reflect the fact that our atmosphere has been oxygen-rich for two billion years.

Impressive though this history may be, why should we consider oxygen as a cosmic sign of life? After all, our own planet apparently had life for well over a billion years before life made our atmosphere rich in oxygen molecules. The key to oxygen's current availability lies in the fact that oxygen enters the atmosphere as the result of photosynthesis, which we may call, with only slight exaggeration, nature's way of making starlight come alive. On Earth, photosynthesis uses light as the energy source that converts water (H_2O) and carbon-dioxide (CO_2) molecules into large complex organic molecules, made mostly of carbon, hydrogen, and oxygen atoms, and releases oxygen molecules (O_2) as part of the conversion process. The complex organic molecules store the starlight energy that drives photosynthesis. Plants rely on photosynthesis to grow; animals that eat plants, or that eat other animals that eat plants, likewise draw their energy supply from sunlight energy stored by photosynthesis. We may therefore call all animals "parasites," since they all feast on plant-made food. A second aspect of their parasitism appears in their respiratory processes, which use the oxygen released into the air by plant photosynthesis.

Let us consider the patterns we discussed in chapter 2 and ask what role oxygen might play on other planets as life appears there. We expect that carbon dioxide should be present in any planet's atmosphere, because CO_2 molecules are relatively strong and do not interact with other molecules as readily as oxygen molecules do. If life on another planet uses water as a solvent, then so long as its star shines, photosynthesis should eventually appear, not in the precise modes we find on Earth, but as the basic process of using starlight energy to combine carbon-dioxide and water molecules. Evolutionary selection should then, as it has on Earth, favor anything that can profit from the products of photosynthesis, which store energy for whatever organisms can use it. If a solvent other than water predominates on another planet, we may still expect that in many cases, something like photosynthesis will allow the storage of starlight; so long as the solvents consist largely of hydrogen, carbon, and oxygen, these processes are also likely to release oxygen molecules. Any means that life evolves to store starlight energy in compounds made of hydrogen, carbon, nitrogen, and oxygen atoms will involve the release of oxygen molecules. Hence the presence of large amounts of oxygen in a planet's atmosphere signals the existence of life, and not of ancient life either, but life existing today (or at any rate recently in astronomical terms). Oxygen molecules combine with other molecular types in the slow processes of oxidation, so a planet that does not replenish its atmospheric oxygen supply will lose oxygen from its atmosphere almost completely after a few thousand years have passed.

This does not mean that every planet with life must have oxygen in its atmosphere. Photosynthesis can occur with sulfur playing the role of oxygen; apparently just this occurred during life's early history in volcanic regions where hydrogen sulfide was abundant. Sulfur-based photosynthesis may have come first, with the structures that had evolved to make photosynthesis more feasible already in place as the changeover to oxygen photosynthesis occurred. On a planet still rich in hydrogen sulfide, sulfur photosynthesis might remain dominant. This would seem to imply that we ought to search for sulfur as well as oxygen in our quest to find life. However, we already know that sulfur occurs in many situations seemingly devoid of life, such as volcanoes on the Earth and on Io. What makes oxygen molecules so useful in the search for life consists in the supposition that many forms of life may have developed water-based photosynthesis, combined with the fact that oxygen provides a nearly unambiguous sign of life, because no good explanation other than life exists for large amounts of oxygen in a planet's atmosphere. We may therefore miss some planets with life by searching for oxygen, but if we find it, we may shout with some confidence, Here it is, life itself.

Even better than finding oxygen alone would be the discovery of oxygen and methane (CH_4). In an oxygen-rich environment, methane combines with oxygen so quickly that a few years will suffice to remove all detectable amounts of methane from the atmosphere unless methane is steadily replenished. Methane provides about one part in a million of Earth's atmosphere, produced in metabolic processes by anaerobic bacteria in estuaries, marshes, rice paddies, and ruminant animals. A planet whose atmosphere shows both large amounts of oxygen and detectable amounts of methane would rank at the top of the list of planets thought to have life. Note that methane by itself proves nothing; many planets, such as the sun's gas giants, have plenty of it. Only when methane appears in an oxygen-rich situation can we suspect—almost conclude—that life exists.

Methane plus oxygen would be an excellent indication, but perhaps we should not ask for everything. Most of those who search for life will be delighted if they ever find planets with significant amounts of oxygen in their atmospheres. "We want to search for oxygen," Alain Léger says. "The question is, how do we do it?"

First, find your planet. As the bulk of this book has demonstrated, finding Earth-like planets will not be easy, but it should occur within the next few decades at the latest. Once you have your Earth-like planet, arrange for your telescope to receive light reflected by the planet from its star in amounts sufficient to allow detailed spectroscopic analysis. (This is a daunting task indeed, since the star outshines the planet by a factor of several million in the infrared and several billion in the visible-light portions of the spectrum.) Study the spectrum of the reflected starlight carefully, seeking absorption lines produced by oxygen. Unfortunately, oxygen molecules (O_2) do not produce easily observable lines; fortunately, ozone (O_3) molecules do. Ozone molecules arise after ultraviolet radiation from a star breaks oxygen molecules apart: Some of the individual oxygen atoms combine with oxygen molecules to form triplets. Their characteristic absorption lines appear at relatively long infrared wavelengths, close to 10^{-3} centimeter. An ideal system to search for both oxygen and ozone would examine infrared wavelengths that bracket this range with sufficient sensitivity for observers to detect even modest amounts of ozone in a planet's atmosphere. No scientist will disagree with the principle that ozone implies the existence of oxygen molecules. If you find ozone and therefore establish the existence of oxygen, you still must convince skeptics that life exists, but you will have little trouble persuading them to invest in further research to settle the issue.

Dreaming of identifying planets with ozone, Alain Léger and his fellow enthusiasts envision the Darwin mission, an interferometer system equipped with a sensitive spectrometer and sent to nearly Jupiter's distance from the sun to avoid the interference from the zodiacal light. Though Darwin's conception actually preceded that of the NASA interferometer, by now the two systems occupy about the same point in development: They are fine ideas to which scientists are attempting to fit numbers, but far from the funding state. The name Darwin refers to the famous biologist and does not represent an acronym, although we might easily invent a name in appropriately acronymic form, such as Deep-space Astronomical Research With Interferometric Networks, or, if we prefer French, Développement Astrophysique des Réseaux Warrantés pour l'Interférometrie Non-terrestre. Darwin's "cheap" version (more precisely, the "not so immensely expensive for a pure-science project" version) would employ two telescopes, each with a 1.3-meter mirror, held 10 meters apart; the full mission would use four 0.9-meter telescopes in a cross pattern with 30-meter arms.

For now, the details of the proposed Darwin system, and how they might differ from what we have named the *FEAST* system, matter far less than the news that scientists and administrators have come to take both projects seriously. This marks a sea-change in attitude during the mid-1990s, one that began before the new planets were discovered in 1995 and 1996, and one that seems likely to continue to rise in public and governmental attention as we sail toward the new millennium. No one knows what Darwin or *FEAST* will cost or whether we can achieve their goals by the year 2025. I predict that NASA and ESA will either combine their efforts or see neither Darwin nor *FEAST* in our lifetimes, for neither the American nor the European public alone will support such an expensive, long-duration effort (the trip to Jupiter itself will require 3 to 5 years) in competition with another, possibly more successful one.

What everyone should know beyond any issue of prediction is that our long isolation as a planetary system has ended. We now know what many astronomers, professional and amateur, suspected for a long time—that a multitude of stars in the Milky Way have planets. Among these new worlds, or ones not far distant or different from them, we may find life, thus fulfilling the conclusions that the Roman poet Lucretius reached two thousand years ago. "For since infinite space stretches out on all sides, and atoms of numberless number and incalculable quantity fly about in all directions quickened by eternal movement," Lucretius wrote,

> it can in no way be considered likely that this is the only heaven and earth created, and all those other atoms there beyond are doing nothing. For this world was created by Nature after atoms had collided spontaneously and at random in a thousand ways, driven together blindly, uselessly, without any results, when at last suddenly the particular ones combined which could become the perpetual starting points of things we know—earth, sea, sky, and the various kinds of living things. Therefore, we must acknowledge that such combinations of other atoms happened elsewhere in the universe to make worlds such as this one, held in the close embrace of the aether.

In the early eighteenth century, Alexander Pope wrote of "worlds unnumber'd"—worlds that humans have now started to enter in catalogues for closer examination. Astronomers like myself have understandable prejudices, but I here assert, without fear of immediate contradiction, that from a perspective of five hundred years or so, our descendants will mark the end of the twentieth century as the time when humanity found the new worlds that will forever change our attitudes toward the cosmos.

This globular star cluster, Omega Centauri, packs several million stars into a region that spans only 70 light years, 15 times the distance from the sun to its closest neighbor stars.

Further Reading

Angel, Roger, and Neville Woolf. "Searching for Life on Other Planets." *Scientific American* 274, pp. 60–66 (April 1996).

Cohen, Nathan. *Gravity's Lens*. New York: John Wiley & Sons, 1988.

Davies, Paul. *Are We Alone*? New York: Basic Books, 1995.

Dole, Stephen, and Isaac Asimov. *Planets for Man*. New York: Random House, 1964.

Drake, Frank, and Dava Sobel. *Is Anyone Out There? The Scientific Search for Extraterrestrial Intelligence*. New York: Delacorte Press, 1992.

Goldsmith, Donald. *The Hunt for Life on Mars*. New York: Dutton, 1997.

———. *Einstein's Greatest Blunder? The Cosmological Constant and Other Fudge Factors in the Physics of the Universe*. Cambridge, Mass.: Harvard University Press, 1995.

———, ed. *The Quest for Extraterrestrial Life*. Mill Valley, Calif.: University Science Books, 1980.

Goldsmith, Donald, and Tobias Owen. *The Search for Life in the Universe*. 2d ed. Reading, Mass.: Addison-Wesley, 1992.

Hoyle, Fred. *The Black Cloud*. New York: Signet Books, 1957.

Klass, Philip. *UFOs Explained*. New York: Vintage Books, 1976.

———. *UFOs: The Public Deceived*. Buffalo: Prometheus Books, 1983.

———. *UFO Abductions: A Dangerous Game*. Buffalo: Prometheus Books, 1989.

"Life in the Universe." Special issue of *Scientific American* (September 1994).

Mermin, David. *Space and Time in Special Relativity*. New York: McGraw-Hill, 1972.

Morrison, David, and Tobias Owen. *The Planetary System*. 2d ed. Reading, Mass.: Addison-Wesley, 1992.

Regis, Edward, Jr., ed. *Extraterrestrials: Science and Alien Intelligence*. Cambridge: Cambridge University Press, 1985.

Sagan, Carl. *The Demon-Haunted World: Science as a Candle in the Dark*. New York: Random House, 1996.

———. *Pale Blue Dot*. New York: Random House, 1994.

Thorne, Kip. *Black Holes and Time Warps: Einstein's Outrageous Legacy*. New York: W. W. Norton & Co., 1994.

Credits

Frontispiece - Lick Observatory photograph

Figures 1 and 2 - Photographs by Donald Goldsmith

Chapter 4 Opener - Lick Observatory photograph

Chapter 5 Opener - Lick Observatory photograph

Chapter 6 Opener - Photograph from National Optical Astronomy Observatories and Kitt Peak National Observatory

Figures 3 and 4 - Courtesy of Philip J. Klass

Page 216 - National Optical Astronomy Observatories photograph

Color Plate 7 - NASA/Space Telescope Science Institute

Color Plate 9 - National Radio Astronomy Observatory

Color Plate 10 - Max-Planck-Institut für Aeronomie and European Space Agency

Color Plate 11 - NASA/Space Telescope Science Institute

Color Plate 13 - Courtesy of Drs. Bradford Smith & Richard Terrile and JPL/NASA

Color Plate 15 - NASA

Color Plate 17 - NASA

Color Plate 18 (top) - NASA

Color Plate 18 (bottom) - NASA

Color Plate 19 (top) - NASA

Color Plate 19 (bottom) - NASA

Color Plate 24 (top) - Courtesy of Profs. Roger Angel & Neville Woolf and the University of Arizona

Color Plate 24 (bottom) - Courtesy of Dr. Charles Beichman and JPL/NASA

Glossary

Absolute brightness: The intrinsic brightness or luminosity of an object.

Absolute temperature scale: Temperature measured on a scale (denoted K) that begins at absolute zero and increases by the same units as those in the Celsius (Centigrade) system, so that water freezes at 273.16 K and boils at 373.16 K.

Absolute zero: The lowest point on any temperature scale, the temperature at which all motion ceases, except for certain quantum-mechanical effects. Absolute zero occurs at 0 K, at −273.16 C, and at −459.67 F.

Absorption line: A limited region of the electromagnetic spectrum within which the intensity of radiation falls significantly below that of neighboring spectral regions.

Acceleration: A change in an object's velocity, which may consist of a change in its speed, its direction of motion, or both.

Accretion: A relatively gradual addition of matter that adds to the mass of an object.

Angular resolution: Ability to see details clearly, expressed as the smallest angular size separating two objects that can be distinguished as separate sources of light.

Angular separation: The distance on the sky between two objects, measured in degrees, minutes, or seconds of arc.

Angular size: The fraction of a circle (360 degrees) over which an object appears to extend as an observer sees it, measured in degrees, minutes of arc (each 1⁄60 of a degree), and seconds of arc (each 1⁄60 of a minute of arc).

Apparent brightness: The brightness that an object appears to have as an observer measures it, hence a brightness that depends both on the object's absolute brightness and on the distance of the observer from the object.

Asteroid: One of the small objects, made mainly of rock or of rock and metal, that orbit the sun, mainly between the orbits of Mars and Jupiter, and range in size from 600 miles in diameter down to objects less than a few hundred yards across.

Astrometry: Measurement of the precise positions and motions of stars.

Astronomical Unit: The average distance from the Earth to the sun, equal to 149,597,900 kilometers, or 92,955,000 miles, and abbreviated as A.U.

Astrophysics: The physics of astronomical events.

Atom: The smallest electrically neutral unit of an element, consisting of a nucleus made of one or more protons and zero or more neutrons, around which orbit a number of electrons equal to the number of protons in the nucleus.

Binary star system: Two stars relatively close to one another that orbit around their common center of mass.

Black hole: An object with such enormous gravitational force at its surface that nothing, not even light, can escape from it.

Brown dwarf: An object with too little mass to become a star, because the slow contraction that heats its interior never raises the temperature to the point that nuclear fusion begins.

Callisto: The outermost of Jupiter's Galilean satellites.

Carbon cycle: A series of nuclear-fusion reactions that converts hydrogen into helium nuclei, using carbon nuclei as the site on which these reactions begin.

Center of mass: The point within an object or group of objects from which the quantity (mass × distance) is the same in any two opposite directions.

Centigrade (Celsius) temperature scale: A scale of temperature that registers the freezing point of water at and the boiling point of water at 100 degrees.

Chondrite: A meteorite that contains inclusions called chondrules.

Chondrule: A small, round granule of matter embedded within some meteorites.

Comet: A fragment of primitive solar-system material, a "dirty snowball" made of ice, rock, dust, and frozen carbon dioxide (dry ice), typically with an orbit around the sun much larger than any planet's orbit.

Conservation of angular momentum: A statement of the fact that an object that experiences no net force will maintain a constant angular momentum. Since the angular momentum is proportional to the object's mass times its rate of spin times the square of its size, this fact implies that objects that experience no net force will spin more rapidly as they contract.

Coronagraph: An instrument that uses an opaque mask to block the light from a bright object, typically the sun, in order to reveal much fainter objects nearby.

Coronagraphic finger: A projection that holds an opaque mask in front of a bright object.

Darwin mission: A mission proposed to the European Space Agency to search for extrasolar planets with an interferometer sent to about Jupiter's distance from the sun.

Degenerate matter: Matter affected in its bulk motions by the "exclusion principle," a rule of quantum mechanics that limits the number of particles that can have nearly the same location and the same velocity.

Doppler effect: The change in the frequencies and wavelengths of photons arriving from a source that has a motion along the observer's line of sight, either of approach or of recession. Astronomers routinely use this effect to determine the velocities of objects toward or away from the Earth by measuring these changes, called Doppler shifts, for light waves.

Drake Equation: An equation, first derived by Frank Drake, that summarizes the factors that lead to an estimate of the number of civilizations that now exist in the Milky Way.

Doppler shift: The amount of the changes produced by the Doppler effect.

Dry ice: Frozen carbon dioxide (CO_2).

Eccentricity: A measure of the noncircularity of an ellipse. If the ratio of the ellipse's short and long axes equals $\mathbf{q}$, the eccentricity $\mathbf{e}$ equals the square root of $(1-\mathbf{q}^2)$. Thus for a circle, with $\mathbf{q}=1$, $\mathbf{e}=0$.

Electromagnetic radiation: Streams of photons.

Electron: An elementary particle with one unit of negative charge, which orbits the nucleus of an atom.

Element: The set of all atomic nuclei that have the same number of protons in the nucleus.

Ellipse: A closed curve defined by the property that the sum of the distances from any point on the curve to two fixed points called "foci" remains constant. If the two foci coincide, the ellipse becomes a circle.

Emission line: A narrow region of the spectrum at which especially large numbers of photons appear within a small range of frequencies and wavelengths.

Europa: One of Jupiter's Galilean satellites, intriguing for its covering of water ice, which may conceal liquid water.

Evolution: In biology, the result of the process of "natural selection" (differential success at reproduction), which under certain circumstances causes groups of similar organisms, called species, to change over time so that their descendants differ significantly in structure and appearance.

ExNPS: An acronym for NASA's project "Exploration of Neighboring Planetary Systems."

Extrasolar: Pertaining to objects beyond the solar system; extrasolar planets are planets that orbit stars other than the sun.

Force: The capacity to do work or to cause a physical change; an influence that tends to produce an acceleration in the direction of its application.

Frequency: Of photons, the number of oscillations per second.

Galaxy: A large group of stars, typically numbering in the hundreds of millions up to hundreds of billions, and usually containing significant amounts of gas and dust, held together by the mutual gravitational attraction among the stars.

Galilean satellites: Jupiter's four largest satellites, discovered in 1610 by Galileo.

***Galileo* spacecraft:** The spacecraft sent by NASA to Jupiter in 1990, which arrived in December 1995, dropped a probe into Jupiter's atmosphere, and continued to orbit the giant planet, photographing it and its Galilean satellites.

Gamma rays: Photons with the highest frequencies and shortest wavelengths.

Ganymede: The largest of Jupiter's satellites and the largest satellite in the solar system, a bit larger than the planet Mercury.

Giant planet: A planet similar to Jupiter, Saturn, Uranus, or Neptune, consisting primarily of hydrogen and helium, plus a solid core of rock and ice, with a mass ranging from a dozen or so Earth masses up to many hundred or even a few thousand times the mass of Earth.

Gravitational lensing: The bending of light rays passing close by a massive object by the object's gravitational force, which produces a distorted view of the source of light, often producing long, thin arcs of light and multiple images of a single source.

Gravitational microlensing: Gravitational lensing in cases where the distortion of an image cannot be observed, but the changes in apparent brightness caused by the lensing effect can be measured.

Greenhouse effect: The trapping of infrared radiation by a planet's atmosphere, which raises the temperature on and immediately above the planet's surface.

Habitable zone: The region surrounding a star, a spherical shell bounded by inner and outer spherical surfaces, within which the star's heat can maintain one or more potential solvents in the liquid state.

Helium: The second-lightest and second most abundant element, whose nuclei all contain two protons and either one or two neutrons.

Hydrogen: The lightest and most abundant element, whose nuclei all contain one proton and either no neutron or one neutron.

Infrared: Electromagnetic radiation consisting of photons whose wavelengths are all slightly longer, and whose frequencies are all slightly lower, than the photons that form visible light.

Inner planets: The sun's planets Mercury, Venus, Earth, and Mars, all of which are small, dense, and rocky in comparison with the giant planets.

Inorganic: Not involving life or the chemistry on which life is based; in particular, not based on carbon atoms.

Interferometer: A combination of two or more photon detectors that achieves the angular resolution of a single, much larger detector whose diameter equals the spacing between the most widely separated components of the interferometer.

Interferometry: The science of observing small angular sizes with interferometers.

Interplanetary dust: Dust spread among the planets in the solar system and in other planetary systems. The density of interplanetary dust is much greater than that of interstellar dust and rises to especially high values close to the central star.

Interstellar dust: Dust particles, each made of a million or so atoms, probably ejected into interstellar space from the atmospheres of highly extended stars.

Io: The innermost of Jupiter's Galilean satellites, notable for the active volcanoes on its surface that spew sodium-laden compounds to create a transient atmosphere.

Kelvin temperature scale: The absolute scale of temperatures, whose units (abbreviated K) are the same as those for the Celsius scale, in which water freezes at 273.16 degrees and boils at 373.16 degrees.

Kepler Project: A proposed means of detecting extrasolar planets by the transit method.

Kilogram: A basic unit of mass in the metric system, containing one thousand grams. On the Earth's surface, one kilogram has a weight of approximately 2.2 pounds.

Kilometer: A unit of length in the metric system, equal to one thousand meters and approximately 0.62137 mile.

Kinetic energy: Energy associated with motion.

Lagrangian point: A point where a small object can maintain a stable position under the gravitational influence of two much more massive objects. In the case of a planet orbiting a star, two of the Lagrangian points lie on the planet's orbit, one of them one-sixth of an orbit ahead of the planet and the other one one-sixth of an orbit behind the planet as it moves along its orbital path.

Light: Photons whose frequencies and wavelengths fall within the band denoted as visible light, between infrared and ultraviolet.

Light year: The distance that light travels in one year, equal to about six trillion miles.

Luminosity: Of an object, the total amount of energy emitted each second in photons of all types.

Macho: An acronym for "massive compact halo object," detected by gravitational microlensing but of unknown form and composition.

Major axis: The long axis of an ellipse.

Metallic hydrogen: A high-pressure phase of hydrogen, either liquid or solid, in which hydrogen conducts electric currents well.

Meteor: A luminous streak of light produced by the heating of a meteoroid as it passes through Earth's atmosphere.

Meteorite: A meteoroid that survives its passage through Earth's atmosphere.

Meteoroid: An object of rock or metal, or a metal-rock mixture, orbiting the sun, smaller than an asteroid or a planet.

Meter: The fundamental unit of length in the metric system, equal to approximately 39.37 inches.

Microwaves: Photons whose frequencies and wavelengths place them between radio and infrared in the spectrum of electromagnetic radiation.

Milky Way: The galaxy that contains the sun and approximately 300 billion other stars.

Minor axis: The short axis of an ellipse.

Molecule: A stable grouping of two or more atoms.

Neap tides: Tides of the lowest range in Earth's oceans, which occur near first-quarter and last-quarter moon, when the sun's tide-raising influence tends to counteract that of the moon.

Neutron: An elementary particle with no electric charge, stable when part of an atomic nucleus but subject to rapid decay when isolated.

Neutron star: A tremendously dense object, the core of an exploded star, typically about a dozen miles across, in which almost all the protons and electrons have combined to form neutrons, and the object has become essentially a single giant nucleus that consists almost entirely of neutrons.

Newton's law of gravitation: Newton's statement of the fact that any two objects with mass attract one another with an amount of force that varies in proportion to the product of the object's masses, divided by the square of the distance between their centers.

Newton's laws of motion: Newton's description of how objects react to net forces upon them, stating that 1) an object with no net force will undergo no acceleration; 2) the amount of acceleration **a** in response to a net force **F** equals **F/m**, where **m** is the object's mass; 3) if one object exerts a force on another, the second object exerts a force in the opposite direction, equal in amount, on the first.

Nuclear fusion: The joining of two nuclei under the influence of strong forces, which occurs only if the nuclei approach one another to a distance approximately the size of a proton, about 0.4×10^{-13}inch.

Nucleus: The central region of an atom, composed of one or more protons and zero or more neutrons.

Occultation: The passage of one object in front of another, completely blocking the light from the farther object.

Organic: Referring to chemical compounds containing carbon atoms as an important structural element; carbon-based molecules. Also, having properties associated with life.

Ozone: Molecules made of three oxygen atoms (O_3). Ozone molecules high in Earth's atmosphere shield the surface against most ultraviolet radiation.

Parsec: A unit of length equal to about 3.262 light years.

Planet: An object in orbit around a star that is not another star or a brown dwarf and has a size at least as large as Pluto.

Planetesimal: An object much smaller than a planet, capable of building planets through numerous mutual collisions.

Plate tectonics: Slow motions of plates of the crust of Earth and similar planets.

Proper motion: A star's apparent motion against the background of much more distant stars (not the result of the Earth's motion around the sun, which must be accounted for to find the proper motion), arising from the star's velocity with respect to the sun as both stars orbit the center of the Milky Way.

Proton: An elementary particle with one unit of electric charge, and one of the two basic components of an atomic nucleus.

Proton-proton cycle: A series of nuclear-fusion reactions that changes four protons into one helium nucleus, with the production of new kinetic energy. Most stars produce most of their energy through the proton-proton cycle.

Protoplanet: A planet during its later formation stages.

Protoplanetary disk: The disk of gas and dust that surrounds a star, especially during the earliest part of the star's lifetime, from and within which planets may form.

Protostar: A star in formation, contracting from a much larger cloud of gas and dust under its self-gravitation.

Protosun: The sun during its formation process, which ended 4.6 billion years ago.

Pulsar: An object that emits pulses of electromagnetic radiation at regularly spaced intervals, apparently a rapidly rotating neutron star.

Quasar: A "quasi-stellar radio source," an object almost starlike in appearance but in fact billions of light years from the Milky Way. Quasars have enormous luminosities and are notable sources not only of radio but also of infrared and visible light.

Radio: Electromagnetic radiation with long wavelengths and low frequencies.

Radio astronomy: The subfield of astronomy that specializes in detecting and analyzing sources of radio emission.

Red dwarf: A star with a considerably lower mass and luminosity than the sun's.

Reflex effect: The acceleration of an object in response to a force upon it; in particular, a star's acceleration in response to a planet's gravitational force, which causes the star to make a small orbit around the center of mass of the star-planet system while the planet makes a much larger orbit.

Runaway greenhouse effect: A greenhouse effect that builds on itself when the heating of a planet's surface increases the rate of evaporation, which in turn increases the greenhouse effect.

Satellite: A relatively small object that orbits a much larger and more massive one; more precisely, both objects orbit their common center of mass.

Self-gravitation: The gravitational force that parts of an object exert on all the other parts.

SETI: The Search for Extraterrestrial Intelligence.

Shooting star: A popular name for a meteor.

Solar nebula: The protoplanetary disk that surrounded the protosun.

Solar system: The sun plus the objects that orbit the sun, including nine planets, their satellites, asteroids, meteoroids, comets, and interplanetary dust.

Solvent: A liquid capable of dissolving another substance; a liquid in which molecules can float and interact.

Spectral line: A narrow region of the spectrum that shows either a noticeably large number of photons (an emission line) or a markedly small number (an absorption line) in comparison to neighboring regions.

Spectrometer: An instrument for careful observation and measurement of a spectrum.

Spectrum: The distribution of photons by frequency or wavelength, often shown as a graph of the number of photons with each particular frequency or wavelength.

Spring tides: The highest and lowest tides during a month, which occur close to full moon and new moon, when the sun's tide-raising influence combines with the moon's.

Standard model: Of planet formation, the generally accepted scenario in which a rotating cloud of gas and dust flattened as it contracted, producing a disk within which ice and rock planetesimals formed and then collided to make planets, while the central mass continued its contraction and became a star.

Star: A mass of gas held together by self-gravitation, in whose center nuclear-fusion reactions produce kinetic energy that heats the entire star, causing its surface to glow.

Stellar wind: Streams of particles, mostly electrons, protons, and helium nuclei, expelled from the outer layers of a star into interstellar space.

Strong forces: One of the four basic types of forces (others are gravitational, electromagnetic, and weak forces), always attractive, which operate only at extremely small distances and bind together protons and neutrons in the nuclei of atoms.

Submillimeter radiation: Photons with wavelengths between 0.1 and 1 millimeter, a type of microwaves.

Supernova: A star that explodes at the end of its nuclear-fusing lifetime, expelling its outer layers and shining with an apparent brightness comparable to that of a small galaxy for several weeks.

Synchroton radiation or synchroton emission: Photons emitted when electrically charged particles moving at nearly the speed of light accelerate in the presence of a magnetic field.

Temperature: The measure of the average kinetic energy of random motion within a group of particles. On the absolute or Kelvin temperature scale, the temperature is directly proportional to the average kinetic energy per particle.

Tides: Bulges that arise in a deformable object as the result of the differences in the amounts of gravitational force that a nearby object produces at different points.

Titan: Saturn's large satellite, almost as large as Ganymede, and the only solar-system satellite with a thick atmosphere, made mostly of nitrogen molecules (as is the Earth's).

Transit: The passage of one object in front of a more distant one that does not completely cover the more distant object.

UFOs: Unidentified flying objects; objects seen in the skies of Earth to which no natural explanation can be easily assigned.

Ultraviolet: Photons with frequencies and wavelengths intermediate between those of visible light and X rays, hence with frequencies somewhat greater than, and wavelengths somewhat less than, those of visible light.

Velocity component: Part of an object's total velocity, singled out for the direction toward which the component points. For example, the velocity of an object moving in orbit has a component that points along the line of sight to an observer, and another component perpendicular to that line of sight.

Velocity resolution: The ability of an observational instrument to measure velocities. These measurements therefore have an associated observational error in velocity equal to the velocity resolution.

Visible light: Photons with wavelengths and frequencies detectable by the human eye.

***Voyager* spacecraft:** The NASA spacecraft, named *Voyager 1* and *Voyager 2*, which were launched from Earth in 1978 and passed by Jupiter and Saturn a few years later. *Voyager 2* proceeded to pass by Uranus in 1986 and Neptune in 1989.

Wavelength: The distance between successive wave crests or wave troughs in any regular wave motion.

White dwarf: The core of a former star, which stands exposed after the star has lost its outer layers and ceased nuclear fusion and consists mainly of carbon nuclei and electrons. The white dwarf continues to radiate stored energy but generates no new kinetic energy because no nuclear fusion occurs within it.

X rays: Photons with frequencies greater than those of ultraviolet but less than those of gamma rays.

Zodiac: The band of 12 constellations that circles the sky as seen from Earth, through which the sun, moon, and planets appear to move during the course of a year. These motions occur within this limited portion of the sky because the planets all orbit the sun in nearly the same plane, and the moon's orbit around the Earth likewise lies close to this plane.

Zodiacal cloud: The swarm of dust particles orbiting the sun, most of which concentrate toward the plane of the Earth's orbit, which is close to the plane of all the other planets' orbits as well.

Zodiacal light: The faint glow visible shortly before dawn or soon after dusk that concentrates in the zodiacal constellations and arises from the reflection of sunlight by myriad dust particles.

Index

F

G

H

I

J

K